Ontology Engineering

Synthesis Lectures on the Semantic Web: Theory and Technology

Editors

Ying Ding, *Indiana University*
Paul Groth, *University of Amsterdam*

Founding Editor

James Hendler, *Rensselaer Polytechnic Institute*

Synthesis Lectures on the Semantic Web: Theory and Technology is edited by Ying Ding of Indiana University and Paul Groth of University of Amsterdam. Whether you call it the Semantic Web, Linked Data, or Web 3.0, a new generation of Web technologies is offering major advances in the evolution of the World Wide Web. As the first generation of this technology transitions out of the laboratory, new research is exploring how the growing Web of Data will change our world. While topics such as ontology-building and logics remain vital, new areas such as the use of semantics in Web search, the linking and use of open data on the Web, and future applications that will be supported by these technologies are becoming important research areas in their own right. Whether they be scientists, engineers or practitioners, Web users increasingly need to understand not just the new technologies of the Semantic Web, but to understand the principles by which those technologies work, and the best practices for assembling systems that integrate the different languages, resources, and functionalities that will be important in keeping the Web the rapidly expanding, and constantly changing, information space that has changed our lives.

Topics to be included:

- Semantic Web Principles from linked-data to ontology design

- Key Semantic Web technologies and algorithms

- Semantic Search and language technologies

- The Emerging "Web of Data" and its use in industry, government and university applications

- Trust, Social networking and collaboration technologies for the Semantic Web

- The economics of Semantic Web application adoption and use

- Publishing and Science on the Semantic Web

- Semantic Web in health care and life sciences

Ontology Engineering
Elisa F. Kendall and Deborah L. McGuinness

ISBN: 978-3-031-79485-8 print
ISBN: 978-3-031-79486-5 ebook

DOI 10.1007/978-3-031-79486-5

A Publication in the Springer series
SYNTHESIS LECTURES ON THE SEMANTIC WEB: THEORY AND TECHNOLOGY
Lecture #18

Series Editors: Ying Ding, Indiana University, Paul Groth, University of Amsterdam
Founding Editor: James Hendler

Series ISSN 2160-4711 Print 2160-472X Electronic

Ontology Engineering

Elisa F. Kendall
Thematix Partners LLC

Deborah L. McGuinness
Rensselaer Polytechnic Institute

SYNTHESIS LECTURES ON THE SEMANTIC WEB: THEORY AND TECHNOLOGY #18

ABSTRACT

Ontologies have become increasingly important as the use of knowledge graphs, machine learning, natural language processing (NLP), and the amount of data generated on a daily basis has exploded. As of 2014, 90% of the data in the digital universe had been generated in the preceding two years, and the volume of data was projected to grow from 3.2 zettabytes to 40 zettabytes in the following six years. The very real issues that government, research, and commercial organizations are facing in order to sift through this amount of information to support decision-making alone mandate increasing automation. Yet, the data profiling, NLP, and learning algorithms that are ground-zero for data integration, manipulation, and search provide less-than-satisfactory results unless they utilize terms with unambiguous semantics, such as those found in ontologies and well-formed rule sets. Ontologies can provide a rich "schema" for the knowledge graphs underlying these technologies as well as the terminological and semantic basis for dramatic improvements in results. Many ontology projects fail, however, due at least in part to a lack of discipline in the development process. This book, motivated by the Ontology 101 tutorial given for many years at what was originally the Semantic Technology Conference (SemTech) and then later from a semester-long university class, is designed to provide the foundations for ontology engineering. The book can serve as a course textbook or a primer for all those interested in ontologies.

KEYWORDS

ontology; ontology development; ontology engineering; knowledge representation and reasoning; knowledge graphs; Web Ontology Language; OWL; linked data; terminology work

Contents

Foreword by Dean Allemang

When we think of the history of engineering in computing, we see a repeating trend. We start by identifying a discipline, say, computer programming. Then as we write more and more programs and discover issues around maintenance and reuse of the programs, we come to understand that there is a strategic view that we can take, and we start to have a repeatable engineering discipline, like software engineering. As we recognize that the sort of strategic work that engineers do, beyond software, is a real and tangible thing, we give it a name, usually a name that includes the word "architecture." We have seen this pattern with business modeling/architecture/engineering, data modeling/architecture, enterprise architecture, and so on. There is a classic progression from some tactical practice to a strategic awareness. As our understanding of effective action in a field matures, we develop specialized language, advanced tools, and specific methods for working. Together, these things make effective practice in the field a repeatable, shareable activity.

"Knowledge engineering," as a buzzword, has been around for about as long as any of the other computer engineering words—perhaps longer than some more recent words like "enterprise architecture." But it has also been the most controversial. When I was in graduate school, I continually had to defend the fact that I was studying "knowledge" when my peers were doing more topical studies like networking, databases, or memory management. Around that time, Alan Newell postulated that a system could be productively described at the "knowledge level," but this postulate remained controversial for many years. The vast majority of software was built without paying attention to entities at the "knowledge level," paying more attention to programming languages, memory management, and a new discipline that gathered around the name, "software engineering."

That was over three decades ago, and today, the value of a "knowledge graph" (in contrast to a database) is now well accepted. The huge search engines manage massive amounts of information connected in highly contextualized ways. We now accept that knowledge is key for managing massively distributed shared data (i.e., on the Web), and that Ontologies play a central role in representing, modularizing, and distributing that knowledge.

So, it is timely that we now see a volume that makes the construction and distribution of ontologies into an engineering discipline. What heretofore resembled an artistic endeavor, performed with idiosyncratic methods only by uniquely skilled artisans, is now becoming a repeatable engineering practice. The volume you hold in your hands represents a watershed in this field. As a reader of this book, you are now part of the first generation of ontology engineers.

It should come as no surprise that such a seminal book on ontology engineering is also a book about software engineering. And it is about business architecture. Mathematics and logic. Library science. Search engine optimization. Data modeling. Software design and methodology. Ontology engineering is not a hodge-podge of techniques taken from all these fields; by its very nature, ontology engineering genuinely and essentially touches all facets of computer science application. Earlier disciplines could draw boundaries; business architecture is distinct from data modeling. Data modeling is distinct from processing. Enterprise architecture is different from content management. Each of these fields has a clear boundary, of what they do and do not have to deal with. The ontology engineer, by contrast, has to be conversant in all of these fields, since the job of the ontology is to bring the system together and connect it to its operation in the real world of business.

As you read this book, you will understand that ontology engineering is not just the sum of all these parts, but it is indeed a new activity of its own, with its own goals and challenges. This book breaks this activity down into its basic pieces, providing goals, methodologies, tools, and insights at each level.

You couldn't ask for better authors for such a work. In those early days when we debated the nature and even the existence of something called "knowledge," McGuinness was right there in the thick and thin of the global discussion. She is a seasoned researcher and educator who has been central to the development of our understanding of knowledge management, representation, or processing throughout the long and varied history of knowledge-based systems. Whenever anyone in the history of knowledge management has considered an alternative to any representational or processing aspect of knowledge, McGuinness was there, and can tell today's student where every line of argument will lead.

Kendall has spent years (decades?) in the trenches, in the field, in a variety of industries, behind the hyper-secure firewalls of the military and in the open data universe of global standards, applying these ideas even as they were still being worked out. She's been there through thick and thin, as the computing world has developed its understanding of the role of knowledge in its systems. While others were talking about how one might do these things, she was out there doing them. It's about time she writes this down for the rest of us.

Wherever you start your journey—as a beginning student with an interest in building knowledge-based systems that make use of the rich contextual information in today's world, or as a seasoned veteran in one of the related fields—this work will show you a context in which all these things come together to form something new. Now it's time to take your first step.

Dean Allemang
Principal Consultant
Working Ontologist LLC

Foreword by Richard Mark Soley, Ph.D.

Recently, I was speaking at a conference and spent most of the hour talking about metadata and semantics. At the end of the hour, we had some time for audience questions and answers, so I opened the floor to questions. The first question floored me: "Great speech, Dr. Soley, but I'm still hazy on one point. What do you mean by metadata?"

Great speech, indeed! I was grumpy and tired and probably hungry too. "Three!" said I, immediately drawing a reply from the confused audience member.

"Three what?" he asked.

"Congratulations," I answered. "You've just reinvented metadata!"

This vignette repeats time and time again across the world as metadata, semantics and ontology swirl around every discussion of data value. "Data is the new oil!" we're told; but it's not. What's important is what that data means in context. And the context is the metadata, or the (unfortunately) implied ontology governing the meaning of that data.

Defining those contexts is not a trivial thing; if it was, one of the myriad attempts over the years to define the "semantics of everything" would likely have worked. Instead of redefining Yet Another Middleware solution (as often is the case, starting with an easy or solved problem first), we'd have a way to easily connect any two or more software applications. Natural language translation would be a snap. User interfaces would be obvious!

But that hasn't happened, and it likely won't. Semantics of information are highly dependent on context (vertical market, application usage, time of day, you name it). Corralling data into usable information remains hard but worth the trouble. No longer will governments publish all their collected data without explaining what it means; something that has already happened!

At the Object Management Group, ontologies are of supreme importance. This three-decade-old well-established standards organization, having gone through middleware and modeling phases, is now tightly focused on vertical markets; more than three-quarters of all standards currently in development are focused on vertical markets like financial systems, retail point-of-sale systems, military command-and-control systems, manufacturing systems, and the like. The core of all of these standards are ontologies that bring orderly semantics to the syntax of the connections. And high-quality design and engineering of ontologies allows them to withstand the changing vicissitudes of markets and gives some hope that ontological (semantic) information might cross domains.

Well-engineered ontologies are therefore the cornerstone of high-quality standards. Far more than mere data models or interface definitions, an ontology leads to both; that is, if you get the semantics right, it is much more likely that your interface definitions, database metamodels—in fact, all of the artifacts that you need will almost design themselves. Some or all of the necessary artifacts forming the basis of good programming may simply "fall out" of the ontology!

I hope this gets you thinking about how to engineer a high-quality ontology that stands the test of time. You're ready for an explanation of exactly how to do that.

Richard Mark Soley, Ph.D.
Chairman and Chief Executive Officer
Object Management Group, Inc.

Preface

Several years ago, when Jim Hendler first suggested that we contribute our Ontology 101 tutorial from the Semantic Technologies Conference (fondly known as SemTech) in the form of a book to this series, we were convinced that we could crank it out in a matter of weeks or a few months at most. The tutorial was presented as a half-day workshop, and we had nine years' experience in presenting and updating it in response to audience feedback. We knew from feedback that the tutorial itself was truly a firehose, and that making it available in an extended, more consumable and referenceable form would be helpful. We also knew that despite the growing number of books about semantic technologies, knowledge representation and description logics, graph databases, machine learning, natural language processing, and other related areas, there was really very little that provided a disciplined methodology for developing an ontology aimed at long-lived use and reuse. Despite the number of years that have gone by since we began working on it, that sentiment hasn't changed.

The tutorial was initially motivated by the Ontology 101 (Noy and McGuinness, 2001) paper, which was based on an expansion of a pedagogical example and ontology that McGuinness provided for wine and food pairing as an introduction to conceptual modeling along with a methodology for working with description logics (Brachman et al., 1991a). It was also influenced by a number of later papers such as Nardi and Brachman's introductory chapter in the *DL Handbook* (Baader et al., 2003), which described how to build an ontology starting from scratch. None of the existing references, however, really discussed the more holistic approach we take, including how to capture requirements, develop terminology and definitions, or iteratively refine the terms, definitions, and relationships between them with subject matter experts through the development process. There were other resources that described use case development or terminology work, several of which we reference, but did not touch on the nuances needed specifically for ontology design. There were many references for performing some of these tasks related to data modeling, but not for developing an ontology using a data model as a starting point, what distinguished one from the other, or why that mattered. And nothing we found addressed requirements and methodologies for selecting ontologies that might be reused as a part of a new development activity, which is essential today. Nothing provided a comprehensive, end-to-end view of the ontology development, deployment, and maintenance lifecycle, either.

In 2015, we extended the tutorial to a full 13-week graduate course, which we teach together at Rensselaer Polytechnic Institute (RPI), where Dr. McGuinness is a constellation chair and professor of computer and cognitive science. We needed a reference we could use for that course as well as for the professional training that we often provide as consultants. That increased our motivation to put this together, although business commitments and health challenges slowed us down a bit. The content included in this initial edition reflects the original tutorial and the first five lectures of our Ontology Engineering course at RPI. It covers the background, requirements gathering, terminology development, and initial conceptual modeling aspects of the overall ontology engineering lifecycle. Although most of our work leverages the World Wide Web Consortium (W3C) Resource Description Framework (RDF), Web Ontology Language (OWL), SPARQL, and other Semantic Web standards, we've steered away from presenting many technical, and especially syntactic, details of those languages, aside from illustrating specific points. Other references we cite, especially some publications in this series as well as the *Semantic Web for the Working Ontologist* (Allemang and Hendler, 2011), cover those topics well. We have also intentionally limited our coverage of description logic as the underlying technology as many resources exist. The examples we've given come from a small number of use cases that are representative of what we see in many of our projects, but that tend to be more accessible to our students than some of the more technical, domain-specific ontologies we develop on a regular basis.

This book is written primarily for an advanced undergraduate or beginning graduate student, or anyone interested in developing enterprise data systems using knowledge representation and semantic technologies. It is not directed at a seasoned practitioner in an enterprise per se, but such a person should find it useful to fill in gaps with respect to background knowledge, methodology, and best practices in knowledge representation.

We purposefully pay more attention to history, research, and fundamentals than a book targeted for a corporate audience would do. Readers should have a basic understanding of software engineering principles, such as knowing the difference between programs and data, the basics of data management, the differences between a data dictionary and data schema, and the basics of querying a database. We also assume that readers have heard of representation formats including XML and have some idea of what systems design and architecture entail. Our goal is to introduce the discipline of ontology engineering, which relates to all of these things but is a unique discipline in its own right. We will outline the basic steps involved in any ontology engineering project, along with how to avoid a number of common pitfalls, what kinds of tools are useful at each step, and how to structure the work towards a successful outcome.

Readers may consider reading the entire book as a part of their exploration of knowledge engineering generally, or may choose to read individual chapters that, for the most part, are relatively self-contained. For example, many have already used Chapter 3 along with the use case template provided in our class and book materials. Others have found the terminology chapter and related template useful for establishing common vocabularies, enterprise glossaries, and other artifacts independently of the modeling activities that follow.

Our intent is to continue adding chapters and appendices in subsequent editions to support our teaching activities and based on feedback from students and colleagues. We plan to incorporate our experience in ontology engineering over the entire development lifecycle as well as cover patterns specific to certain kinds of applications. Any feedback on what we have presented here or on areas for potential expansion, as we revise and augment the content for future audiences, would be gratefully appreciated.

CHAPTER 1

Foundations

Ontologies have become increasingly important as the use of knowledge graphs, machine learning, and natural language processing (NLP), and the amount of data generated on a daily basis has exploded. Many ontology projects have failed, however, due at least in part to a lack of discipline in the development process. This book is designed to address that, by outlining a proven methodology for the work of ontology engineering based on the combined experience of the authors, our colleagues, and students. Our intent for this chapter is to provide a very basic introduction to knowledge representation and a bit of context for the material that follows in subsequent chapters.

1.1 BACKGROUND AND DEFINITIONS

Most mature engineering fields have some set of authoritative definitions that practitioners know and depend on. Having common definitions makes it much easier to talk about the discipline and allows us to communicate with one another precisely and reliably about our work. Knowing the language makes you part of the club.

We hear many overlapping and sometimes confusing definitions for "ontology," partly because the knowledge representation (KR) field is still maturing from a commercial perspective, and partly because of its cross-disciplinary nature. Many professional ontologists have background and training in fields including formal philosophy, cognitive science, computational linguistics, data and information architecture, software engineering, artificial intelligence, or library science. As commercial awareness about linked data and ontologies has increased, people knowledgeable about a domain but not trained in any of these areas have started to build ontologies for use in applications as well. Typically, these individuals are domain experts looking for solutions to something their IT departments haven't delivered, or they are enterprise architects who have run into brick walls attempting to use more traditional technologies to address tough problems. The result is that there is not as much consensus about what people mean when they talk about ontologies as one might think, and people often talk past one another without realizing that they are doing so.

There are a number of well-known definitions and quotes in the knowledge representation field that practitioners often cite, and we list a few here to provide common grounding:

> *"An ontology is a specification of a conceptualization."* (Gruber, 1993)

This definition is one of the earliest and most cited definitions for *ontology* with respect to artificial intelligence. While it may seem a bit academic, we believe that by the time you finish reading this book, you'll understand what it means and how to use it. It is, in fact, the most terse and most precise definition of ontology that we have encountered. Having said this, some people may find a more operational definition helpful:

> *"An ontology is a formal, explicit description of concepts in a domain of discourse (classes (sometimes called concepts)), properties of each concept describing various features and attributes of the concept (slots (sometimes called roles or properties)), and restrictions on slots (facets (sometimes called role restrictions))."* (Noy and McGuinness, 2001)

The most common term for the discipline of ontology engineering is "knowledge engineering," as defined by John Sowa years ago:

> *"Knowledge engineering is the application of logic and ontology to the task of building computable models of some domain for some purpose."* (Sowa, 1999)

Any knowledge engineering activity absolutely must be grounded in a domain and must be driven by requirements. We will repeat this theme throughout the book and hope that the *"of some domain for some purpose"* part of John's definition will compel our readers to specify the context and use cases for every ontology project you undertake. Examples of what we mean by context and use cases will be scattered throughout the sections that follow, and will be covered in depth in Chapter 3.

Here are a few other classic definitions and quotes that may be useful as we consider how to model knowledge and then reason with that knowledge:

> *"Artificial Intelligence can be viewed as the study of intelligent behavior achieved through computational means. Knowledge Representation then is the part of AI that is concerned with how an agent uses what it knows in deciding what to do."* (Brachman and Levesque, 2004)

> *"Knowledge representation means that knowledge is formalized in a symbolic form, that is, to find a symbolic expression that can be interpreted."* (Klein and Methlie, 1995)

> *"The task of classifying all the words of language, or what's the same thing, all the ideas that seek expression, is the most stupendous of logical tasks. Anybody but the most accomplished logician must break down in it utterly; and even for the strongest man, it is the severest possible tax on the logical equipment and faculty."* (Charles Sanders Peirce, in a letter to editor B. E. Smith of the *Century Dictionary*)

Our own definition of ontology is based on applied experience over the last 25–30 years of working in the field, and stems from a combination of cognitive science, computer science, enterprise architecture, and formal linguistics perspectives.

An ontology specifies a rich description of the

- *terminology, concepts, nomenclature;*

- *relationships among and between concepts and individuals ; and*

- *sentences distinguishing concepts, refining definitions and relationships (constraints, restrictions, regular expressions)*

relevant to a particular domain or area of interest.

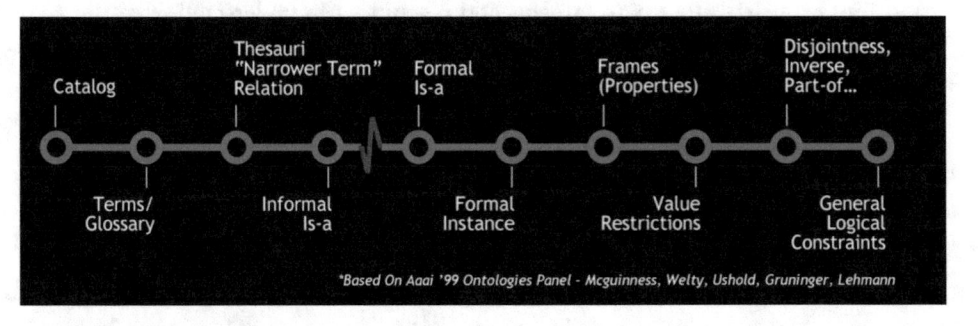

Figure 1.1: Ontology definition and expressivity spectrum.

Figure 1.1 provides an abbreviated view of what we, and many colleagues, call the "ontology spectrum"—the range of models of information that practitioners commonly refer to as ontologies. It covers models that may be as simple as an acronym list, index, catalog, or glossary, or as expressive as a set of micro theories supporting sophisticated analytics.

The spectrum was developed during preparation for a panel discussion in 1999 at an Association for the Advancement of Artificial Intelligence (AAAI) conference, where a number of well-known researchers in the field attempted to arrive at a consensus on a definition of ontology. This spectrum is described in detail in McGuinness, *Ontologies Come of Age* (2003). We believe that an ontology can add value when defined at any level along the spectrum, which is usually determined by business or application requirements. Most of the ontologies we have developed, whether conceptual or application oriented, include at least a formal "is-a" or subclass hierarchy, and often additional expressions, such as restrictions on the number or type of values for a property, (i.e., they fall to the right of the red "squiggle" in the diagram).

Regardless of the level of expressivity and whether the ontology is conceptual in nature or application focused, we expect that an ontology will be: (1) encoded formally in a declarative knowledge representation language; (2) syntactically well-formed for the language, as verified by an appropriate syntax checker or parser; (3) logically consistent, as verified by a language-appropriate reasoner or theorem prover; and (4) will meet business or application requirements as demonstrated through extensive testing. The process of evaluating and testing an ontology is both science and art, with increasingly sophisticated methods available in commercial tools, but because no "one size fits all," we typically need multiple tools to fully vet most ontologies. We will discuss some of the more practical and more readily available approaches to ontology evaluation in later chapters of this book.

1.2   LOGIC AND ONTOLOGICAL COMMITMENT

The primary reason for developing an ontology is to make the meaning of a set of concepts, terms, and relationships explicit, so that both humans and machines can understand what those concepts mean. The level of precision, breadth, depth, and expressivity encoded in a given ontology depends on the application: search applications over linked data tend to require broader ontologies and tolerate less precision than those that support data interoperability; some machine learning and natural language processing applications require more depth than others. Ontologies that are intended to be used as business vocabularies or to support data governance and interoperability require more metadata, including clearly stated definitions, provenance, and pedigree, as well as explanatory notes and other usage information than machine learning applications may need. The foundation for the machine-interpretable aspects of knowledge representation lies in a combination of set theory and formal logic. The basis for the metadata stems from library science and terminology work, which we discuss in Chapter 4.

Most people who are interested in knowledge representation took a course in logic at some point, either from a philosophical, mathematical, or linguistics perspective. Many of us also have basic knowledge of set theory, and can draw Venn diagrams showing set intersection when needed, but a little refresher may be helpful.

Logic can be more difficult to read than English, but is clearly more precise:

(forall ((x FloweringPlant))

  (exists ((y Bloom)(z BloomColor))(and (hasPart x y)(hasCharacteristic y z))))

Translation: *Every flowering plant has a bloom which is a part of it, and which has a characteristic bloom color.*

Language: *Common Logic, CLIF syntax* (ISO/IEC 24707:2018, 2018)

Logic is a simple language with few basic symbols. The level of detail depends on the choice of predicates made by the ontologist (e.g., *FloweringPlant*, *hasPart*, *hasCharacteristic*, in the logic, above); these predicates represent an ontology of the relevant concepts in the domain.

1.3   ONTOLOGY-BASED CAPABILITIES

An ontology defines the vocabulary that may be used to specify queries and assertions for use by independently developed resources, processes, and applications. "Ontological commitments are agreements to use a shared vocabulary in a coherent and consistent manner."[1] Agreements can be specified as formal ontologies, or ontologies with additional rules, to enforce the policies stated in those agreements. The meaning of the concepts included in the agreements can be defined precisely and unambiguously, sufficient to support machine interpretation of the assertions. By composing or mapping the terms contained in the ontologies, independently developed systems can work together to share information and processes consistently and accurately.

Through precise definitions of terms, ontologies enable shared understanding in conversations among agents to collect, process, fuse, and exchange information. For example, ontologies can be used to improve search accuracy through query expansion to clarify the search context. Typically, search accuracy includes both precision and recall, meaning that correct query results are returned and relevant answers are not missing. Ontologies designed for information sharing may be used in a number of ways, including but not limited to:

- on their own as terminologies or common vocabularies to assist in communications within and across groups of people;

- to codify, extend, and improve flexibility in XML[2] and/or RDF Schema-based[3] agreements;

- for information organization, for example for websites that are designed to support search engine optimization (SEO) and/ or those that use mark-up per schema.org;[4] or

- to describe resources in a content management system, for example for archival, corporate website management, or for scientific experimentation and reuse.

Ontologies that describe information resources, processes, or applications are frequently designed to support question answering, either through traditional query languages such as SQL[5] or SPARQL,[6] or through business rules, including rule languages such as RuleML,[7] Jess,[8] Flora-2,[9] and commercial production rule languages. They may also be designed to support more complex applications, including:

---

[1] http://www.ksl.stanford.edu/kst/what-is-an-ontology.html.

[2] Extensible Markup Language (XML), see http://www.w3.org/standards/xml/core.

[3] The Resource Description Framework (RDF) Vocabulary Description Language (RDF Schema), available at https://www.w3.org/RDF/.

[4] See https://schema.org/ for more information.

[5] Structured Query Language, see https://docs.microsoft.com/en-us/sql/odbc/reference/structured-query-language-sql?view=sql-server-2017.

[6] SPARQL 1.1 Query Language, available at https://www.w3.org/TR/sparql11-overview/.

[7] The Rule Mark-up Initiative, see http://wiki.ruleml.org/index.php/RuleML_Home.

[8] Jess, the Java Expert System Shell and scripting language, see https://herzberg.ca.sandia.gov/docs/52/.

[9] FLORA-2: Knowledge Representation and Reasoning with Objects, Actions, and Defaults, see http://flora.sourceforge.net/.

- recommender systems, for example, for garden planning, product selection, service provider selection, etc. as part of an event planning system;

- configuration systems such as product configurators or systems engineering design verification and validation;

- policy analysis and enforcement, such as for investment banking compliance and risk management;

- situational analysis systems, such as to understand anomalous behaviors for track and trace, fraud detection, or other business intelligence applications; and

- other complex analyses, such as those required for understanding drug formularies, disease characteristics, human genetics, and individual patient profiles to determine the best therapies for addressing certain diseases.

In other words, ontologies and the technologies that leverage them are well suited to solve problems that are cross-organizational, cross-domain, multi-disciplinary, or that span multiple systems. They are particularly useful in cases where traditional information technologies are insufficiently precise, where flexibility is needed, where there is uncertainty in the information, or where there are rich relationships across processes, systems, and or services that can't be addressed in other ways. Ontologies can connect silos of data, people, places, and things.

In the sections that follow, we will provide examples and modeling patterns that are commonly used to support both lightweight use cases that do not involve much reasoning, as well as richer applications such as recommender systems or systems for policy analysis and enforcement that depend on more representation and reasoning power.

1.4   KNOWLEDGE REPRESENTATION LANGUAGES

Today's approaches to knowledge representation (KR) emerged from 1970s and 1980s research in artificial intelligence, including work in areas of semantic networks, question-answering, neural networks, formal linguistics and natural language processing, theorem proving, and expert systems.

The term knowledge representation is often used to talk about representation of information for consumption by machines, although "good" knowledge representations should also be readable by people. Every KR language has a number of features, most of which are common to software engineering, query, and other languages. They include: (1) a vocabulary, consisting of some set of logical symbols and reserved terms plus variables and constants; (2) a syntax that provides rules for combining the symbols into well-formed expressions; (3) a formal semantics, including a *theory of reference* that determines how the constants and variables are associated with things in the universe of discourse and a *theory of truth* that distinguishes true statements from false ones; and (4) *rules of inference*, that determine how one pattern can be inferred from another. If the logic is *sound*, the rules of inference must preserve truth as determined by the semantics. It is this fourth element, the rules of inference and the ability to infer new information from what we already know, that distinguishes KR languages from others.

Many logic languages and their dialects have been used for KR purposes. They vary from classical first order logic (FOL) in terms of: (1) their syntax; (2) the subsets of FOL they implement (for example, propositional logic without quantifiers, Horn-clause, which excludes disjunctions in conclusions such as Prolog, and terminological or definitional logics, containing additional restrictions); (3) their proof theory, such as monotonic or non-monotonic logic (the latter allows defaults), modal logic, temporal logic, and so forth; and (4) their model theory, which as we mentioned above, determines how expressions in the language are evaluated with respect to some model of the world.

Classical FOL is two-valued (Boolean); a three-valued logic introduces unknowns; four-valued logic introduces inconsistency. Fuzzy logic uses the same notation as FOL but with an infinite range of certainty factors (0.0–1.0). Also, there are differences in terms of the built-in vocabularies of KR languages: basic ISO/IEC 24707:2018 (2018) is a tight, first-order language with little built in terminology, whereas the Web Ontology Language (Bao et al., 2012) includes support for some aspects of set theory.[10]

1.4.1   DESCRIPTION LOGIC LANGUAGES

Description logics (DLs) are a family of logic-based formalisms that represent a subset of first order logic. They were designed to provide a "sweet spot" in that they have a reasonable degree of expressiveness on the ontology spectrum, while not having so much expressive power that it is difficult to build efficient reasoning engines for them. They enable specification of ontologies in terms of concepts (classes), roles (relationships), and individuals (instances).

Description logics are distinguished by (1) the fact that they have a formal semantics, representing decidable fragments of first order logic, and (2) their provisions for inference services, which include sound and complete decision procedures for key problems. By *decidable*, we mean that there are effective algorithms for determining the truth value of the expressions stated in the logic. Description logics are highly optimized to support specific kinds of reasoning for implementation in operational systems.[11]

Example types of applications of description logics include:

---

[10]   For more information on general first-order logics and their use in ontology development, see Sowa (1999) and ISO/IEC 24707:2018 (2018).

[11]   For more information on description logics, KR and reasoning, see Baader et al. (2003) and Brachman and Levesque (2004).

- configuration systems—product configurators, consistency checking, constraint propagation, etc., whose first significant industrial application was called PROSE (McGuinness and Wright, 1998) and used the CLASSIC knowledge representation system, a description logic, developed by AT&T Bell Laboratories in the late 1980s (Borgida et al., 1989);

- question answering and recommendation systems, for suggesting sets of responses or options depending on the nature of the queries; and

- model engineering applications, including those that involve analysis of the ontologies or other kinds of models (systems engineering models, business process models, and so forth) to determine whether or not they meet certain methodological or other design criteria.

1.5   KNOWLEDGE BASES, DATABASES, AND ONTOLOGY

An ontology is a conceptual model of some aspect of a particular universe of discourse (or of a domain of discourse). Typically, ontologies contain only "rarified" or "special" individuals, representing elemental concepts critical to the domain. In other words, they are comprised primarily of concepts, relationships, and axiomatic expressions.

One of the questions that we are often asked is, "*What is the difference between an ontology and a knowledge base?*" Sometimes people refer to the knowledge base as excluding the ontology and only containing the information about individuals along with their metadata, for example, the triples in a triple store without a corresponding schema. In other words, the ontology is separately maintained. In other cases, a knowledge base is considered to include both the ontology and the individuals (i.e., the triples in the case of a Semantic Web-based store). The ontology provides the schema and rules for interpretation of the individuals, facts, and other rules comprising the domain knowledge.

A knowledge graph typically contains both the ontology and related data. In practice, we have found that it is important to keep the ontology and data as separate resources, especially during development. The practice of maintaining them separately but combining them in knowledge graphs and/or applications makes them easier to maintain. Once established, ontologies tend to evolve slowly, whereas the data on which applications depend may be highly volatile. Data for well-known code sets, which might change less frequently than some data sets, can be managed in the form of "OWL ontologies," but, even in these cases, the individuals should be separate from the ontology defining them to aid in testing, debugging, and integration with other code sets. These data resources are not ontologies in their own right, although they might be identified with their own namespace, etc.

Most inference engines require in-memory deductive databases for efficient reasoning (including commercially available reasoners). The knowledge base may be implemented in a physical, external database, such as a triple store, graph database, or relational database, but reasoning is typically done on a subset (partition) of that knowledge base in memory.

1.6   REASONING, TRUTH MAINTENANCE, AND NEGATION

Reasoning is the mechanism by which the logical assertions made in an ontology and related knowledge base are evaluated by an inference engine. For the purposes of this discussion, a logical assertion is simply an explicit statement that declares that a certain premise is true. A collection of logical assertions, taken together, form a logical theory. A consistent theory is one that does not contain any logical contradictions. This means that there is at least one interpretation of the theory in which all of the assertions are *provably true*. Reasoning is used to check for contradictions in a collection of assertions. It can also provide a way of finding information that is implicit in what has been stated. In classical logic, the validity of a particular conclusion is retained even if new information is asserted in the knowledge base. This may change if some of the prior knowledge, or preconditions, are actually hypothetical assumptions that are invalidated by the new information. The same idea applies for arbitrary actions—new information can make preconditions invalid.

Reasoners work by using the rules of inference to look for the "deductive closure" of the information they are given. They take the explicit statements and the rules of inference and apply those rules to the explicit statements until there are no more inferences they can make. In other words, they find any information that is implicit among the explicit statements. For example, from the following statement about flowering plants, if it has been asserted that x is a flowering plant, then a reasoner can infer that x has a bloom y, and that y has a characteristic which includes a bloom color z:

> *(forall ((x FloweringPlant))*
>> *(exists ((y Bloom)(z BloomColor))(and (hasPart x y)(hasCharacteristic y z))))*

During the reasoning process, the reasoner looks for additional information that it can infer and checks to see if what it believes is consistent. Additionally, since it is generally applying rules of inference, it also checks to make sure it is not in an infinite loop. When some kind of logical inconsistency is uncovered, then the reasoner must determine, from a given invalid statement, whether or not others are also invalid. The process associated with tracking the threads that support determining which statements are invalid is called *truth maintenance*. Understanding the impact of how truth maintenance is handled is extremely important when evaluating the appropriateness of a particular inference engine for a given task.

If all new information asserted in a knowledge base is monotonic, then all prior conclusions *will*, by definition, remain valid. Complications can arise, however, if new information negates a prior statement. "Non-monotonic" logical systems are logics in which the introduction of new axioms can invalidate old theorems (McDermott and Doyle, 1980). What is important to understand when selecting an inference engine is whether or not you need to be able to invalidate previous assertions, and if so, how the conflict detection and resolution is handled. Some questions to consider include the following.

- What happens if conclusive information to prove the assumption is not available?

- The assumption cannot be proven?

- The assumption is not provable using certain methods?

- The assumption is not provable in a fixed amount of time?

The answers to these questions can result in different approaches to negation and differing interpretations by non-monotonic reasoners. Solutions include chronological and "intelligent" backtracking algorithms, heuristics, circumscription algorithms, justification or assumption-based retraction, depending on the reasoner and methods used for truth maintenance.

Two of the most common reasoning methods are forward and backward chaining. Both leverage "if-then" rules, for example, "If it is raining, then the ground is wet." In the forward chaining process, the reasoner attempts to match the"if" portion (or antecedent) of the rule and when a match is found, it asserts the "then" portion (or the consequent) of the rule. Thus, if the reasoner has found the statement "it is raining" in the knowledge base, it can apply the rule above to deduce that "The ground is wet." Forward chaining is viewed as data driven and can be used to draw all of the conclusions one can deduce from an initial state and a set of inference rules if a reasoner executes all of the rules whose antecedents are matched in the knowledge base.

Backward chaining works in the other direction. It is often viewed as goal directed. Suppose that the goal is to determine whether or not the ground is wet. A backward chaining approach would look to see if the statement, "the ground is wet," matches any of the consequents of the rules, and if so, determine if the antecedent is in the knowledge base currently, or if there is a way to deduce the antecedent of the rule. Thus, if a backward reasoner was trying to determine if the ground was wet and it had the rule above, it would look to see if it had been told that it is raining or if it could infer (using other rules) that it is raining.

Another type of reasoning, called tableau (sometimes tableaux) reasoning, is based on a technique that checks for satisfiability of a finite set of formulas. The semantic tableaux was introduced in 1950s for classical logic and was adopted as the reasoning paradigm in description logics starting in the late 1990s. The tableau method is a formal proof procedure that uses a refutation approach—it begins from an opposing point of view. Thus, when the reasoner is trying to prove that something is true, it begins with an assertion that it is false and attempts to establish whether this is satisfiable. In our running example, if it is trying to prove that the ground is wet, it will assert that it is NOT the case that the ground is wet, and then work to determine if there is an inconsistency. While this may be counterintuitive, in that the reasoner proposes the opposite of what it is trying to prove, this method has proven to be very efficient for description logic processing in particular, and most description logic-based systems today use tableau reasoning.

Yet another family of reasoning, called logic programming, begins with a set of sentences in a particular form. Rules are written as clauses such as H :- B1, ... Bn. One reads this as H or the "head" of the rule is true if B1 through Bn are true. B1-Bn is called the body. There are a number of logic programming languages in use today, including Prolog, Answer Set Programming, and Datalog.

1.7   EXPLANATIONS AND PROOF

When a reasoner draws a particular conclusion, many users and applications want to understand why. Primary motivating factors for requiring support for explanations in the reasoners include interoperability, reuse, trust, and debugging in general. Understanding the provenance of the information (i.e., where it came from and when) and results (e.g., what sources were used to produce the result, what part of the ontology and rules were used) is crucial to analysis. It is essential to know which information sources contributed what to your results, particularly for reconcilliation and understanding when there are multiple sources involved and those sources of information differ. Most large companies have multiple databases, for example, containing customer and account information. In some cases there will be a "master" or "golden" source, with other databases considered either derivative or "not as golden"—meaning, that the data in those source databases is not as reliable. If information comes from outside of an organization, reliability will depend on the publisher and the recency of the content, among other factors.

Some of the kinds of provenance information that have proven most important for interpreting and using the information inferred by the reasoner include:

- identifying the information sources that were used (source);

- understanding how recently they were updated (currency);

- having an idea regarding how reliable these sources are (authoritativeness); and

- knowing whether the information was directly available or derived, and if derived, how (method of reasoning).

The methods used to explain why a reasoner reached a particular conclusion include explanation generation and proof specification. We will provide guidance in some depth on metadata to support provenance, and on explanations in general in the chapters that follow.

CHAPTER 2

Before You Begin

In this chapter we provide an introduction to domain analysis and conceptual modeling, discuss some of the methods used to evaluate ontologies for reusability and fit for purpose, identify some common patterns, and give some high-level analysis considerations for language selection when starting a knowledge representation project.

2.1 DOMAIN ANALYSIS

Domain analysis involves the systematic development of a model of some area of interest for a particular purpose. The analysis process, including the specific methodology and level of effort, depends on the context of the work, including the requirements and use cases relevant to the project, as well as the target set of deliverables. Typical approaches range from brainstorming and high-level diagramming, such as mind mapping, to detailed, collaborative knowledge and information modeling supported by extensive testing for more formal knowledge engineering projects. The tools that people use for this purpose are equally diverse, from free or inexpensive brainstorming tools to sophisticated ontology and software model development environments. The most common capabilities of these kinds of tools include:

- "drawing a picture" that includes concepts and relationships between them, and

- producing sharable artifacts, that vary depending on the tool—often including web sharable drawings.

Analysis for ontology development leverages domain analysis approaches from several related fields. In a software or data engineering context, domain analysis may involve a review of existing software, repositories, and services to find commonality and to develop a higher-level model for use in re-engineering or to facilitate integration (de Champeaux, Lea, and Faure, 1993; Kang et al., 1990). In an artificial intelligence and knowledge representation context, the focus is on defining structural concepts, their organization into taxonomies, developing individual instances of these concepts, and determining key inferences for subsumption and classification for example, as in Brachman et al. (1991b) and Borgida and Brachman (2003). From a business architecture perspective, domain analysis may result in a model that provides wider context for process re-engineering, including the identification of core competencies, value streams, and critical challenges of an organization, resulting in a common view of its capabilities for various purposes (Ulrich and McWhorter, 2011). In library and information science (LIS), domain analysis involves studying a broad range of information related to the knowledge domain, with an aim of organizing that knowledge as appropriate for the discourse community (Hjørland and Albrechtsen, 1995). Domain analysis to support ontology development takes inspiration from all of the above, starting from the knowledge representation community viewpoint and leveraging aspects of each of the others as well as from the terminology community (ISO 704:2009, 2009).

The fact that the techniques we use are cross-disciplinary makes the work easier for people from any of these communities to recognize aspects of it and dive in. At the same time, this cross-disciplinary nature may make the work more difficult to understand and master, involving unfamiliar and sometimes counterintuitive methods for practitioners coming from a specific perspective and experience base. Some of the most common disconnects occur when software or data engineers make assumptions about representation of relationships between concepts, which are first class citizens in ontologies, but not in some other modeling paradigms such as entity relationship diagramming (Chen, 1976) or the Unified Modeling Language, Version 2.5.1 (2017). Java programmers, for example, sometimes have difficulty understanding inheritance—some programmers take short cuts, collecting attributes into a class and "inheriting from it" or reusing it when those attributes are needed, which may not result in a true is-a hierarchy. Subject matter experts and analysis who are not fluent in logic or the behavior of inference engines may make other mistakes initially in encoding. Typically, they discover that something isn't quite right because the results obtained after querying or reasoning over some set of model constructs are not what they expected. Although there may be many reasons for this, at the end of the day, the reasoners and query engines only act as instructed. Often the remedy involves modeling concepts and relationships more carefully from the domain or business perspective, rather than from a technical view that reflects a given set of technologies, databases, tagging systems, or software language.

2.2   MODELING AND LEVELS OF ABSTRACTION

Sometimes it helps people who are new to knowledge representation to provide a high-level view of where ontologies typically "play" in a more traditional modeling strategy. Figure 2.1, below, provides a notional view of a layered modeling architecture, from the most abstract at the highest level to very concrete at the lowest. This sort of layering is common to many modeling paradigms.

An ontology can be designed at any level of abstraction, but with reference to Figure 2.1, typically specifies knowledge at the context, conceptual, and/or logical layers. By comparison, entity-relationship models can be designed at a conceptual, logical, or physical layer, and software models at the physical and definition layers. Knowledge bases, content management systems, databases, and other repositories are implemented at the instance layer. In terms of the Zachman Framework™,[12] which is well known in the data engineering community, an ontology is typically specified with

---

[12]   See https://www.zachman.com/about-the-zachman-framework.

respect to at least one of the elements (what, how, where, who, when, why) across the top three rows of the matrix—executive perspective, business perspective, and architect's perspective. A comprehensive ontology architecture might include some or all of these perspectives, with increasing granularity corresponding to increasingly specific levels in the framework.

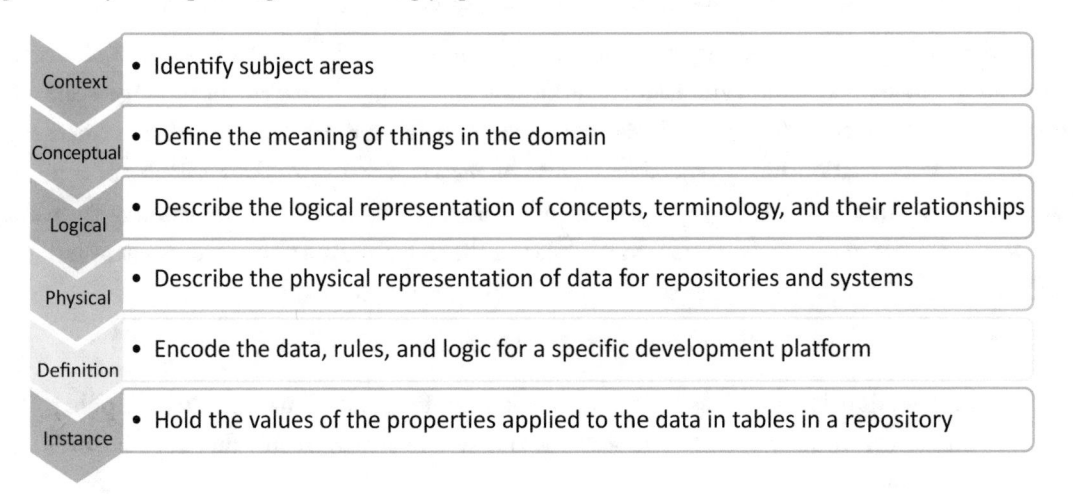

Figure 2.1: Abstraction layers.

An ontology architecture developed to address requirements of a finance application might include:

1. a foundational layer of reusable and customized ontologies for metadata and provenance description (e.g., based on Dublin Core,[13] SKOS (Simple Knowledge Organization System),[14] and the PROV Ontology (PROV-O),[15] potentially with extensions that are context-specific);

2. a domain-independent layer of reusable ontologies covering standard concepts for dates and times, geopolitical entities, languages, and other commonly used concepts, building on the metadata layer—there are standardized and de facto standards that may be used for this purpose;

3. a domain-dependent layer of reusable ontologies covering standards for the domain area relevant to the use cases—any known standard vocabularies should be considered, such as the Financial Industry Business Ontology (FIBO)[16] for financial

---

[13] See http://www.dublincore.org/specifications/.
[14] See https://www.w3.org/standards/techs/skos#w3c_all.
[15] See https://www.w3.org/TR/prov-o/.
[16] More on FIBO can be found at https://spec.edmcouncil.org/ and at https://www.omg.org/industries/finance.htm.

products and services or the Unified Medical Language System (UMLS) for medical applications; and

4. a domain-dependent layer of ontologies specific to the problems of interest, that build on and extend ontologies in the other three layers

Separation of abstraction layers in an ontology architecture is one of a number of strategies that can improve organization and facilitate evolution and change management for large-scale applications. Key questions for an ontologist starting to organize a domain model include whether a given concept or relationship is general enough to be reused, either by other ontologies or applications, and whether that concept or property is independent from others they are working with. The answers can assist in determining how to modularize the domain. At a domain-dependent level an architect might incorporate further separation based on criteria such as

- behavioral, functional, and structural knowledge for engineering systems;

- presentation-layer, process logic, business logic, information infrastructure, and policy infrastructure-related (network, security, and system management) functions in software; and

- "as designed," "as built," "as configured," and "as maintained" for product oriented systems.

Identifying the set of concerns and abstraction layers appropriate for a large-scale ontology development effort should be done early in the process to ensure that the resulting ontologies support these modularization strategies and to avoid downstream rework.

2.3   GENERAL APPROACH TO VOCABULARY DEVELOPMENT

Ontologies can be designed for general use, without any specific application in mind (i.e., as business or scientific vocabularies and controlled vocabularies), or to support applications such as searching and data mining, information, data and systems integration, business intelligence and decision support systems, recommendation systems, question answering systems, natural language processing, predictive analytics, machine learning, or other systems involving automated reasoning and/or rules-based analysis. Use of a logically consistent ontology as the vocabulary used to drive rules and decision management systems increases rule set completeness and accuracy, reduces development time, error, and maintenance costs, and increases user satisfaction with the resulting systems (Nitsche et al., 2014).

Like any other engineering task, the starting point for analysis purposes is to gather requirements and source materials. We use two important tools to support requirements analysis, called Use Cases and Competency Questions. Use cases, which we discuss in detail Chapter 3, provide the structure we need for organizing requirements. Competency questions, which we cover in Section 3.8, are a set of representative questions and example answers that a knowledge graph or application must be able to answer to ensure success. The requirements and use case analysis process ensures sufficient coverage and limits scope creep. Some of the materials that may be relevant, depending on the domain and purpose of the ontology, include not only any available specifications, (e.g., for existing systems or data sources), but policies, procedures, regulations, international and national standards, published dictionaries and glossaries (given that the source is authoritative), and so forth. Source materials should be referenced in the individual use cases that they support and in metadata describing the source(s) used to develop definitions in the ontologies.

If a suitable set of requirements is available to start with, then development of use cases and competency questions, reflecting those requirements, is typically the next step. For example, in financial services, banks and other financial intermediaries need to be able to answer questions raised by their regulators, such as the following.

- What is my current exposure to a given counterparty?

- What is my current exposure with respect to a specific financial instrument (e.g., stock, bond, derivative, future, option) or set of related instruments?

- What is my current exposure to instruments originating from a specific region or country?

Most banks of size, including the *globally systemically important banks* (G-SIBs) and many other smaller institutions, are required to demonstrate fundamental improvements in risk data aggregation and reporting. The requirements for doing so are specified in Basel Committee on Banking Supervision (BCBS) Principles for effective risk data aggregation and risk reporting (BCBS 239[17]). A preliminary analysis would

- include these questions and others derived from the BCBS 239 document in use cases;

- search the laws and regulations in applicable jurisdictions (in this case, in the EU, including BCBS-related regulations, those of the International Accounting Standards Board (IASB), regulations and rules published by the European Central Bank (ECB), and any national or regional equivalents), for glossaries, other definitions, and explanatory content;

- identify the concepts mentioned in the competency questions, such as "counterparty," "exposure," "financial instrument," and "region," as provided in those regulations and other legal dictionaries, including their relationships to one another; and

---

[17] https://www.bis.org/publ/bcbs239.pdf.

- identify the touch points (systems, data sources, policies, procedures, reports, and so forth) where those concepts are used or referenced, both implicitly and explicitly—these form a subset of the actors required for use case development.

Another very early and critical task is to identify key subject matter experts (SMEs) and stakeholders that can validate the competency questions, definitions, critical source materials, and use cases, as well as provide usage scenarios for the relevant terms. The stakeholders and possibly some of the SMEs will also be represented as actors in the use cases. We have customized traditional use case development for ontology and semantically enabled application development purposes, which we will discuss in more detail in Chapter 3.

From the use cases and competency questions, we recommend iterative domain analysis starting with a "thread" that covers the basic content (concepts and relationships), scoped to cover the competency questions developed during use case analysis, which will ground further work and assist in prioritizing what to build out next. An initial analysis should identify the concepts and terminology that are central to the domain or application and the high-level relationships between those concepts. The natural relationships and topic areas identified in the definitions will also suggest how the concepts can be grouped into independent ontologies that reference one another, for example, covering business entities in one ontology, contracts and the definition of a counterparty in another, and financial instruments in yet another, and then further refining those concepts into an increasingly broad and deep network.

Once you have a high-level ontology architecture in mind, and before a great deal of time is spent defining specific concepts, it is important to understand and articulate the architectural trade-offs you're making with your broader team and stakeholders, as well as the technical benefits you believe you will gain from making those trade-offs. The nature of the information and kinds of competency questions that need to be answered, as identified in the use cases, should drive the architecture, approach, ontology scope, and design decisions you make. Documenting and explaining the requirements as you understand them through the use cases, identifying stakeholders and SMEs and getting approval for your interpretation of the requirements together with the set of information sources (actors in your use cases) and reference materials you plan to use is essential at this stage in any project. Understand and communicate the pros and cons of adopting any standard (actual or ad hoc) ontology that you plan to depend on. We recommend that you reuse standardized, well-tested, and available ontologies whenever possible, and a discussion of the trade-offs with respect to such decisions should also be included in your use cases.

Section 2.4 discusses vocabulary development from a more business-oriented perspective, although it is relevant for any situation where requirements are lacking.

2.4   BUSINESS VOCABULARY DEVELOPMENT

In cases without a business or technical specification as a starting point, for example, when the goal is to develop a general-purpose vocabulary for some domain rather than an application-specific ontology, then we suggest that the place to start is with a business architecture (Business Architecture Guild, 2014).[18] A business architecture is essential to development of a general-purpose business vocabulary in the absence of other sources of requirements and scoping information. Use cases and competency questions can be derived from the capability and information maps developed as a part of the business architecture process. The use cases, including usage scenarios and competency questions (again, the set of questions that the use case entails—questions that a repository needs to be able to answer, an application needs to cover, etc.), should weave a path through the capability and information maps, such that every capability and every information element is addressed by at least one use case. The individual capabilities (one or more) should expand to multiple scenarios in the use cases. Information elements should be represented as actors in those use cases. This ensures that the set of competency questions developed provides sufficient coverage of the domain for that business area.

Ontology development can be initiated once a preliminary set of use cases with sufficient domain coverage is available. The target ontology, or family of ontologies, should contain the terms, definitions, and relevant metadata that enable every competency question to be asked and answered. This includes coverage of any ancillary vocabulary identified/used in the use cases, capability maps and descriptions, and information maps and descriptions from the business architecture, as appropriate. Again, depending on the use cases, it may be appropriate to reuse existing ontologies that cover the required terminology, extending them as appropriate. Typically, and especially in scientific domains, there are openly available, and sometimes standard ontologies on which a gap analysis and evaluation for reusability can be performed. In such cases, the ontology development process includes reuse of the selected ontology(ies), mapping terms from different ontologies to each other, and developing extensions to fill any gaps. Test cases, test content, and a set of test competency questions that can later be used for regression testing are also an essential piece of the development process.

Depending on the environment and business requirements, policies and procedures for managing, evolving, and governing the ontology and the development artifacts may also be required. The policies and procedures should cover review cycles, version management, test case and test content development, automation of as much of the testing and review as possible, and management of the interim artifacts that themselves can provide tremendous value to the organization. The use cases, which provide a bridge from a given capability to the vocabulary needed to perform that capability, must cite all reference materials—regulatory and internal policy documents, international

---

[18]   See https://www.businessarchitectureguild.org/default.aspx.

and national standards, other relevant references and sanctioned source materials for definition development—and source and product repositories and artifacts, as well as all stakeholders and other actors, that are appropriate for that use case. These references fill in some of the gaps that the business architecture might not naturally cover.

Deliverables from a business architecture-based ontology engineering process include the business architecture artifacts (value streams, capability maps, information maps, interview questions and responses, gap analysis, and so forth), a set of use cases that leverage the capability and information maps, and related ontology artifacts (e.g., models, diagrams, documentation), and a publishable repository containing the ontologies that comprise the business vocabulary as well as all of the other artifacts, interlinked, for web-based access by the relevant stakeholders.

2.5   EVALUATING ONTOLOGIES

A number of dimensions should be considered when analyzing or classifying ontologies as a part of a development process, for potential reuse, or for other purposes such as refactoring to assist in maintenance activities. Many papers on ontology evaluation cite completeness, correctness, understandability, and usability as the basic measures for determining reusability (Tartir et al., 2005; Duque-Ramos et al., 2011). Metrics describing a particular ontology may reflect the content, inherent structures or patterns, or the semantics of the model itself. Other metrics may indicate the nature of the repositories or other resources that the model describes, or the functionality required of a target application. Understanding these dimensions and how a given ontology stacks up against them can be important in determining whether an ontology is appropriate for reuse. For example, the expressive power required to reason with an ontology will affect the performance as well as reasoning efficiency in any target application. Expressivity, and in fact any of the other dimensions highlighted below, are only interesting if the ontology in question provides *epistemological adequacy* (i.e., the ontology provides sufficient coverage in terms of content richness and scope).

Key semantic aspects[19] to consider include:

- the level of expressivity of the representation language used to specify the ontology (as described by the ontology spectrum presented in Chapter 1);

- the level of complexity (including complexity in the number and nature of axioms in the ontology, as well as the processing complexity required to reason with the ontology, with an understanding of how that complexity may impact performance); and

- the level of granularity in the ontology—explicitly underspecifying the details of the terms may be an appropriate strategy for an ontology with high reuse potential, and

---

[19]   Identified at an Ontology Summit held at the National Institute of Standards and Technology (NIST) in 2007, see http://ontolog.cim3.net/cgi-bin/wiki.pl?OntologySummit2007.

particularly for one intended for use in search applications, with increasing levels of detail in dependent layers in the ontology architecture.

Many of the basic ontology development tools in use today provide an indication of the level of expressivity and complexity of an individual ontology as well as over its imports closure, i.e., the result of importing the ontologies it references into an integrated knowledge graph. The underlying metrics used to determine these indicators can also be helpful in identifying potential performance issues in applications that use the ontology. For example, some more advanced relationships that we will discuss later, such as inverse relations and qualified cardinality restrictions, are known to impact performance, and understanding how prevalent these are in the ontology under evaluation may be an important consideration for reuse. Certain axioms that impact performance can be simulated to minimize any potential impact.[20] What is important is to be able to recognize that there could be an issue and determine when simulation is necessary to reduce the impact on business objectives. Some practical metrics for analyzing the content of an ontology are presented on the BioPortal site,[21] and with respect to a knowledge base that uses it (Tartir et al., 2005). Metrics that support ontology analysis with respect to cohesion as a measure of modularity and reusability are discussed in Yao, Orme, and Etzkorn (2005). Metrics that support suitability for project reuse, evolution, and merging are presented in McGuinness et al. (2000).

A comprehensive analysis of model, application, and other characteristics of ontologies that also provides observations on how these characteristics typically apply to classes of applications is given in (Hart, et al., 2004). Some complementary considerations are given in Das, Wu, and Mc-Guinness (2001). According to Hart et al., functional criterion for ontology evaluation may include:

- the degree of relevance to the problem based on the intended use cases;

- the degree of rigor required to support an application and how well the ontology supports this requirement (i.e., ontologies used for prescriptive purposes have a much higher need for correctness than those that are intended for descriptive uses only); and

- the amount and nature of automation envisioned, and the extent to which the ontology was designed to support the automated reasoning requirements of the target system or application.

Model centric perspectives characterize the ontologies themselves and are concerned with their structure, formalism, and dynamics, including:

- the level of authoritativeness, from the least authoritative, and typically broader, shallower ontologies to most authoritative, narrower, and more deeply defined ontologies;

---

[20] See https://doi.org/10.4018/IJSWIS.2018010101, for example.
[21] https://www.bioontology.org/wiki/Ontology_Metrics.

- the source and nature of structure—from passive (transcendent) structure that originates outside of the system, often from direct development, to active (immanent) structure that emerges from the data or behavior;

- the degree of formality, from informal or primarily taxonomic to quite formal, having rigorously defined types, relationships, and theories or axioms;

- model dynamics, ranging from cases where the ontologies themselves are read-only and static or slow to change, to situations in which the ontologies are fluid in nature and evolving; and

- instance dynamics, ranging from cases where the instances asserted in a knowledge-base that reflects the ontology are read-only to those where the instances are volatile, changing continuously

Application-centric perspectives are concerned with how applications use and manipulate ontologies:

- the degree of control or management oversight, ranging from externally focused ontologies that are public in nature and developed by loosely related communities or even by crowd sourcing to those that are internally focused and entirely controlled by the development organization;

- the nature of application changeability, from static (with periodic updates) to dynamic;

- the nature of application coupling—from loosely coupled to tightly coupled;

- the integration focus, if any, ranging from information integration to application integration; and

- the application lifecycle usage demands on the ontology, at design time, at run-time, or in an operational system.

Aspects supporting ontology design and evolution methodologies include:

- the level of conformance with development policies, including those for naming conventions, documentation and annotation usage, requirements for modularity, test and validation policies, and so on, as required;

- the level of conformance with organizational governance policies; and

- the degree to which the ontology is designed to support collaborative development, change management, and other model evolution expectations;

These characteristics are representative of the kinds of things that people need to think about and prioritize when determining whether or not a particular version of an ontology meets requirements and corporate standards, or whether reusing an existing ontology is appropriate for a given situation. Many of these aspects can also assist project managers in understanding and managing the scope of ontology development projects.

2.6   ONTOLOGY DESIGN PATTERNS

Design patterns have proven to be invaluable across modeling and engineering domains, and research to identify common patterns has been ongoing since the early days of the Semantic Web Best Practices working group.[22] One of the first patterns published by the working group described how to model *part—whole* relationships in Web Ontology Language (OWL).[23] Another pattern published by the working group that is cited frequently describes how to represent *n-ary relations*[24] in OWL; others include how to represent *classes as property values*[25] as well as *specified values and value sets.*[26] Despite language evolution and tremendous progress since then on many technology fronts, these patterns continue to be relevant today.

Building on the work done by Best Practices, there are many papers and conference workshops dedicated to developing an ever-growing catalog of ontology design patterns. Most notably, these include the Ontology Design Patterns wiki created by Aldo Gangemi and Valentina Presutti,[27] which covers general design patterns. A number of websites focused on patterns for domain-specific ontology use are available as well. The important takeaway from this is that a combination of reuse of existing ontologies and applying these kinds of patterns in new ontologies results in higher quality, more maintainable, and more reusable ontologies.

2.7   SELECTING A LANGUAGE

Determining what knowledge representation language to use in a development project is similar to making other technology decisions. While some of the issues highlighted below may seem focused more on applications that use the ontologies rather than on the languages per se, it is important to understand whether the language in question has sufficient tool support to meet project requirements. In addition to organizational requirements, decision criteria may include:

1. the intended use of the ontologies, including domain- and application-specific knowledge representation requirements;

---

[22] The Best Practices working group legacy page is available at https://www.w3.org/2001/sw/BestPractices/.
[23] See https://www.w3.org/2001/sw/BestPractices/OEP/SimplePartWhole/index.html.
[24] See https://www.w3.org/TR/swbp-n-aryRelations/.
[25] See https://www.w3.org/TR/2005/NOTE-swbp-classes-as-values-20050405/.
[26] See https://www.w3.org/TR/2005/NOTE-swbp-specified-values-20050517/.
[27] See http://ontologydesignpatterns.org/wiki/Main_Page.

2. the requirements for the knowledge-based systems that will implement the ontologies, including reasoning requirements, question answering requirements, integration with other systems, etc.;

3. the number and nature of any potential transformations required among processes, resources, services to support semantic mediation, and performance requirements specific to the services that require such transformations;

4. ontology and knowledge representation system alignment, de-confliction, and/or ambiguity resolution requirements;

5. collaborative development requirements, especially those where the need for community participation, such as for vetting definitions, is important; and

6. general performance, sizing, timing requirements of the target environment, other operational considerations, such as those specific to highly distributed environments.

While this list is not exhaustive, it provides a view of some of the factors that architects need to consider as a part of the language and broader application environment selection process. For example, even for ontologies whose primary intended use is as an extensible business vocabulary, as a *reference* for human and perhaps automated systems usage, there may be a requirement for basic calculation support in a deployment environment. We typically use the OWL when it provides enough expressive power and reasoning support. OWL does not provide inherent support for arithmetic, analytical processes, or other computations, however, but can be integrated with rule-based processing or supporting functions, which may be sufficient to fill the gap. Certain financial services and business applications may require support for "negation as failure," which also requires the ontology to be extended through the use of rules if OWL is selected as the underlying ontology language. In cases where there are real-time constraints on processing, such as for handling certain kinds of high-throughput transactions, an ontology may be appropriate as the basis for relational or dimensional schema development and validation, and in out-of-band analysis capabilities (e.g., as part of a related business intelligence application), but may not be the right choice for the underlying transaction processing system.

CHAPTER 3

Requirements and Use Cases

In this chapter we provide an introduction to use cases (Jacobson et al., 1992; Jacobson, Spence, and Bittner, 2011), tailored to ontology design, evaluation, and maintenance and for incremental requirements development for constructing semantically enhanced systems.

Ontology engineering is cross-disciplinary, and may benefit from background in logic, terminology, library science, data science and engineering, business analysis, or software engineering depending on the context and requirements. The reasons for developing an ontology range from the need for a reference vocabulary to something more sophisticated such as support for reasoning in the context of business intelligence, predictive analytics, question answering, recommender systems, or other complex tasks. Because of this diversity in requirements and usage, people who are launching projects using semantics may not think about a full range of ontology-specific requirements. Yet, in order to determine what the ontology needs to provide, in what context, for what purpose, given what data, requirements are essential. It is the requirements that provide the basis for determining the *scope* of work and what it means to be *successful*. Without scoping requirements, ontology projects tend to wander down "interesting but not essential" paths, and do not reach focused completion with clear termination criteria. Without an understanding of the success criteria, *how do you know when you're done*? And, how can you test any revisions to the ontology to ensure that you haven't made changes that will invalidate something users depend on?

We have participated in countless projects where there is pressure to skip the requirements step and jump straight into modeling—"throwing a bunch of classes" into an ontology editor such as Protégé or boxes onto a diagram as a starting place. While someone with a longer term or higher-level view may request requirements, many projects proceed too quickly to modeling without paying enough attention to the requirements step. And there are all kinds of excuses—"oh, this is really research, so there aren't requirements." Another common response is "We need a general-purpose, reference ontology for the entire domain, so there is no way to constrain it." Or—"we'll use the development process to figure it out, because we don't really know what we need." In reality there are always requirements, although you may need to work a bit harder to draw them out of the various stakeholders involved and iterate to obtain a representative set. Ontology development is an engineering activity. This means that although there is certainly art and creativity involved, it's also a discipline. Eliciting requirements and determining scoping boundaries can be daunting, but for every domain, there are sub-domains. For every sub-domain, there are capabilities and topical ways of organizing and prioritizing the work with stakeholders. There must be some way of determining what "successful" means and when "you're done"—at least for a given time frame and for a given "release" of an ontology.

Well-written use cases are critical to scoping any ontology. They become a part of the documentation that should be managed and maintained together with other requirements and design artifacts, so that others can understand the design intent, the criteria for success and completion, and serve as support for making informed judgments as to whether an ontology is fit for reuse. Use cases not only summarize the context and requirements for ontology development, they also help establish agreement on requirements with stakeholders.

We advocate a use case-based approach in part because we have not seen other methods for capturing ontology requirements and scoping questions that are better, but also because we have seen far too many people dive in to development without sufficient knowledge of what they should be modeling, how to evaluate the ontology, and, most importantly, *when to stop modeling work.*

3.1   GETTING STARTED

A basic use case is essentially an ordered set of steps, typically defining the interactions between an actor and a system. A more complete use case includes a collection of possible sequences of interactions between a system and its actors relating to a particular goal or objective. In an ontology-driven or ontology-guided system, ontologies provide the communications fabric for those interactions —the content for those communications. So, from the perspective of an ontologist, a complete collection of use cases should define all known system behavior *relevant to the actors*, but with a focus on the *content* exchanged between them rather than on the detailed communications mechanisms, and with emphasis on the knowledge bases and knowledge-based analysis performed, if any. Any system behaviors that are irrelevant to the actors are extraneous from the ontologist's perspective.

Although this may seem a bit heavy on the requirements at first blush, remember that use case development is iterative, starting with a lightweight description and adding increasing detail over time is expected. If you are working without other sources for requirements, then anticipate that it will be even more iterative, and that the resulting use case may have little in common with where you started.

There are templates available on the Web for specifying use cases, ranging from Alistair Cockburn's site,[28] to suggestions in Jacobson's most recent book (Jacobson, Spence, and Bittner, 2011). We've gathered some of the main points included in these and other recommendations, together with additional requirements for semantically enabled systems and ontology development in an extended template that we recommend for this purpose.[29] You likely will not have all of the

---

[28] https://alistair.cockburn.us/.
[29] Our use case template can be downloaded from the Morgan & Claypool book abstract page, here http://bit.ly/2IpFOGp..

details from the outset, but we recommend filling in what you know, as well as listing any known user stories in the body of the use case as a starting point.

Before we begin, or as a part of the initial requirements gathering process, we need to:

- identify the specific sub-domain, topic area, or organizational perspective to focus on;

- gather relevant content—i.e., any documentation (policies, procedures, standards, other sources of terminology) that we know about in advance;

- identify key stakeholders, principal subject matter SMEs, and consumers of the ontology or work product that use it;

- identify any other known interfaces—systems and data sources—that will participate in the use case;

- sketch out any user stories or usage scenarios for the ontology of which you are aware; and

- seed the use case template with as much information as possible.

Each use case should address the basic capabilities or features needed to accomplish some task, or represent a similar set of questions that certain users would like to be able to ask and answer in some context. In an "agile" or "scrum" development environment, one would seed each use case with some number of user stories[30] that have something in common. The user stories would be collected in a usage scenarios/user stories section of the use case, and preserved for traceability purposes.

Consider the flowering plant example described in Chapter 1. User stories derived from the example might include:

- *I would like to purchase flowering plants that should do well in sunny areas of my garden; and*

- *I am designing a south-facing bed that is partially shaded, and would like drought-tolerant plant suggestions.*

Other user stories that have been important to us include:

- *I would like plants that are not likely to attract and be eaten by deer and that tolerate clay soil; and*

---

[30] https://en.wikipedia.org/wiki/User_story.

- *I need to purchase a new tree that is oak-root fungus resistant to replace one that recently died of that disease. It must be selected from the city of Los Altos preferred tree list and, of those, I would like one that flowers and whose leaves turn in the fall.*

From the financial example in Chapter 2, questions such as the following qualify as user stories if rephrased as statements.

- *What is my current exposure to a given counterparty?*

- *What is my current exposure with respect to a specific financial instrument (e.g., stock, bond, swap agreement, future, option) or set of related instruments?*

- *What is my current exposure to instruments originating from a specific region?*

A more general competency question in this last case might be, "From a set of approved plants for a given setting in a specific region, identify all those that are resistant to certain known diseases and pests and that have certain traits in their phenotype." The ability to map a specific question, such as the one about the Los Altos preferred tree list, to something more general, highlights the power of a well-designed ontology.

The user stories should come from stakeholders in the project rather than from developers (or ontologists) to the degree possible, and only from the latter if they are derived. Derived user stories (e.g., from requirements documents or a business architecture) should be reviewed by the stakeholders for validation and prioritization purposes. If you use a business architecture as the starting point, as discussed in Chapter 2, the user stories should be developed from a combination of the capabilities and interviews, such that every capability is involved in at least one user story,[31] and ultimately in at least one use case.

As helpful as user stories can be on their own, one of the downsides of an agile approach that is *solely* based on user stories is that the overall architecture for the system, the larger usage scenario, and the architecture for the ontology may be lacking. To avoid that pitfall we recommend organizing and mapping the user stories into narratives that cover broader aspects or themes reflecting the desired functionality (i.e., usage scenarios, or mini concepts of operations). This will provide a "thread" that flows through a related set of capabilities, interactions, questions, or functions, as the basis for development of a particular use case, as well as for knitting the use cases together into a more cohesive architecture (Patton and Economy, 2014). The goal is to end up with a set of use cases that cover the desired features of the target system or ontology, and in the case of an ontology, that provide a representative set of competency questions, with sample answers that can be used for testing).

---

[31] User stories are more fully described in Section 1.6.

If your project uses an automated issue and project management system (e.g., Trello[32] (a free general tracking tool), or JIRA,[33] a commercial tool aimed largely at software development projects but can also be used successfully for ontology development), every user story can be entered into that system and be identified for the purposes of requirements traceability. If you are using an even more formal requirements management system, then you should be able to (1) select one or more requirements, and (2) relate each requirement to one or more user stories, to ensure that you have good coverage. Every functional requirement must be covered by at least one user story and every user story must become part of one or more use cases. Ideally, you should already be able to say with some certainty what percentage of requirements coverage the use cases will provide, even before you flesh them out. You should also be able to use the use cases to discover gaps in the requirements and user stories, allowing you to go back to the stakeholders to iterate until coverage is largely complete. A traceability map, from the requirements and user stories to use cases, developed as a part of the requirements analysis process, is often helpful, and may be mandated by project stakeholders, depending on the project and methodology adopted by your organization. The traceability map can also act as an invaluable input for testing purposes. In an agile development environment, some subset of requirements/user stories can be selected for each sprint for a more iterative and incremental development approach.

3.2 GATHERING REFERENCES AND POTENTIALLY REUSABLE ONTOLOGIES

Once you have seeded your initial use case(s) with the details you know, the next step is to start gathering resources you can use as the basis for terms, definitions, and relationships. The terms in the user stories and competency questions will provide guidance and additional content as you build out the use case, but your subject matter experts should be able to provide a list of relevant standards and other references to start with. The goal is to gather the best of a representatives set of :

- controlled vocabularies and nomenclatures already available for the domain;

- any well-known hierarchical and or taxonomic resources that are appropriate;

- standard dictionaries for the domain, such as for financial applications, the International Accounting Standards Board /Financial Accounting Standards Board (IASB/FASB) glossaries, regulatory glossaries, and Barron's dictionaries for finance and business terminology; and

---

[32] https://trello.com.
[33] https://www.atlassian.com/software/jira/.

- reference documents including any relevant international (e.g., ISO, IEEE, OMG, W3C, OASIS, etc.) or national (e.g., ANSI, US government agency glossaries, regulations, policies, and procedures for the domain) standard and de-facto standards,

that might be useful as reference materials, to the degree that they are available for the domain you are working in. Any reference you identify, throughout the analysis process, should be documented in the use case and reviewed with your stakeholders.

For example, from the flowering plants user stories, terms including "flowering plant," "shade," and "drought-tolerant" pop out immediately. There are a number of standards and de facto standards related to flowering plants and the climate and/or location in the landscape that they prefer. In the U.S., the U.S. Department of Agriculture (USDA) publishes a plant-hardiness zone map,[34] and *Sunset Magazine* publishes a well-known garden book, updated frequently, with a climate zone guide to assist gardeners in plant selection.[35] Recent literature in California includes extensive coverage of drought-tolerant landscape suggestions, especially in light of the fact that some water districts offer rebates to property owners for evidence that they have replaced lawns with a drought-tolerant, low water usage garden.[36,37] Similarly, resources such as Landscape Plants rated by deer resistance[38] and for plants that tolerate clay soil[39] can provide plant selection guidance. Depending on the other use cases and the target area the garden application is intended to serve, resources, including botanical and gardening standards (e.g., the International Code of Nomenclature for algae, fungi, and plants (Melbourne Code),[40] Royal Horticultural Society Color Chart[41]), growers catalogs, and other well-known taxonomies such as the Linnaean taxonomy and the Plant List, created by the Royal Botanic Gardens, Kew, and Missouri Botanical Garden, would be appropriate starting points. Local garden societies, national and regional parks, and historic gardens, such as the Filoli[42] and Gamble[43] gardens near Stanford University, may also provide helpful reference materials and links to additional information. Another relevant resource is the IBC Agroportal,[44] which provides descriptions of and links to a number of well-known agriculture-related ontologies and vocabularies.

An increasing number of existing standard and de-facto standard ontologies are publicly available on the Web for most scientific domains, and more recently for some business domains.

---

[34] https://planthardiness.ars.usda.gov/PHZMWeb/.
[35] https://www.sunset.com/garden/climate-zones/climate-zones-intro-us-map.
[36] https://www.sunset.com/garden/earth-friendly/water-rebates-resource-list.
[37] https://www.bhg.com/gardening/plans/easy/drought-tolerant-garden-plan/.
[38] https://njaes.rutgers.edu/deer-resistant-plants/.
[39] https://thesensiblegardener.com/?s=clay+soil+plants.
[40] https://www.iapt-taxon.org/nomen/main.php.
[41] http://www.rhsshop.co.uk/category.aspx?id=10000006.
[42] https://filoli.org/.
[43] https://www.gamblegarden.org/.
[44] http://agroportal.lirmm.fr/.

Again, depending on project requirements, intellectual property challenges, and other organizational criteria, it only makes sense to use and extend existing work, rather than starting from scratch. It makes little sense, for example, to develop a new annotation ontology for use in documenting an ontology if the annotations available in the Dublin Core,[45] SKOS,[46] and the Provenance Ontology[47] are sufficient. There are several ontologies available for representing time (e.g., OWL Time,[48] the time ontologies in the Financial Industry Business Ontology (FIBO[49])), units of measure (such as the QUDV annex of the OMG's SysML[50] standard, or the QUDT[51] ontology), and many other basic ontologies that have been tested in various projects. Reuse of these and other well-known, domain-specific ontologies not only improves the likelihood of success and reduces your development costs, but facilitates reuse of your own work by others. One should always document any reference ontologies identified in the use case as well. Use case development is iterative, so these can be eliminated or augmented as needed, but documenting what you've found, even at the start, will be useful to other team members and colleagues participating in or reviewing the use case.

If there is a gap in the content in the ontologies you have identified, then by all means consider extending them rather than starting from scratch. For example, for general metadata about specifications, and based on the OMG (Object Management Group[52]) processes for developing technical standards, the OMG's Architecture Board[53] has published an ontology for specification metadata[54] that extends Dublin Core and SKOS to meet OMG specification authors' needs. Several OMG specifications have used this metadata vocabulary successfully in published standards,[55] and the OMG has recently redesigned its online specification catalog using lightweight mark-up (using the specification metadata expressed as JSON-LD[56]) on its website.

For scientific and, in particular, biomedical informatics work, there are many well-known ontologies that can be extended to support new or revised applications and research. One of the most well-known collections of ontologies for biomedical informatics is the BioPortal,[57] which includes usage statistics and other metrics that can assist you in determining whether or not a given ontology is considered to be well tested and stable. Another well-known repository for ontologies

---

[45]   http://dublincore.org/.
[46]   https://www.w3.org/2004/02/skos/.
[47]   https://www.w3.org/TR/prov-o/.
[48]   https://www.w3.org/TR/owl-time/.
[49]   https://www.omg.org/spec/EDMC-FIBO/.
[50]   https://www.omg.org/spec/SysML/.
[51]   http://www.qudt.org/.
[52]   https://www.omg.org/.
[53]   https://www.omg.org/about/ab.htm.
[54]   http://www.omg.org/techprocess/ab/SpecificationMetadata.rdf.
[55]   All of the Financial Industry Business Ontology (FIBO) ontologies, the Financial Instrument Global Identifier (FIGI), and several non-financial domain ontologies use it today.
[56]   https://www.w3.org/TR/json-ld/.
[57]   http://bioportal.bioontology.org/.

in the Common Logic community is COLORE.[58] There are also many well used and often well maintained "mid-level" ontologies that have seen extensive reuse such as the SWEET Ontology,[59] which is a general-purpose ontology for science and engineering units initially developed by NASA but now under the governance of the Earth System Information Partners (ESIP) federation. Searching to find reusable ontologies for a specific application or domain may take a little time, but is well worth the result. We feel that it is misguided to build a new ontology without performing a thorough search for existing starting points, in fact.

3.3   A BIT ABOUT TERMINOLOGY

One of the early steps in ontology development is terminology excerption (extracting terms and phrases from a corpus) and preliminary definition development from known reference texts, terminologies, and other information resources. Although we will discuss this further in Chapter 4, it is worth mentioning that increasingly we are using terminology extraction to obtain a candidate-controlled vocabulary. Further techniques may be used to link those terms and phrases to existing entities, thus not only providing a candidate controlled vocabulary but also candidate definitions, thereby providing a starting point ontology. These techniques can also provide terminology for user stories, usage scenarios, and competency question development (questions and answers). Depending on the context for the ontology, terminology extraction can be automated, which saves a significant amount of time, although it may require curation with SMEs. Typically, detailed terminology work is done in parallel with use case iteration, and the two activities "feed each other" synergistically. Use of proper domain terminology from the outset can help ease discussions, especially when soliciting requirements interactively with stakeholders.

3.4   SUMMARIZING THE USE CASE

Once you have seeded your use case template, and have a basic "thread" or narrative in mind, the next step is to summarize the use case. A good use case summary should include statements that:

- provide a short description of the basic requirement(s) the use case is intended to support. Be sure to state the business case (the "why"), including any background knowledge that might be important. If there is an existing business requirements document, provide the reference;

- specify the primary goals of the use case, derived from the narrative you've started to develop. Goals should be expressed in short, declarative sentences—being terse but clear is fine;

---

[58] https://code.google.com/archive/p/colore/.
[59] https://sourceforge.net/projects/sweet-ontology/ and https://sourceforge.net/projects/sweet-ontology/.

- identify any known scoping requirements or boundaries. Be clear about any required capabilities that are *out of scope* for this particular use case or, in general, in addition to those that are in scope;

- provide a description of how things should work, success and failure scenarios, with measures of each, consequences, and identify the primary actor(s);

- list any known pre-conditions and post conditions, including any assumptions about the state of the system or world before and after execution of the use case;

- list all known actors and interfaces—identify primary actors, information sources/repositories, interfaces to existing or new systems, and stakeholders for the use case;

- describe known triggers—conditions that initiate the use case, as well as any scheduled events, or a series of events, situation, activity, etc., that cause "something to happen." Any trigger that affects the flow of the use case should be identified here, including any sensor inputs or alarms; and

- identify any known non-functional requirements, such as performance requirements including any sizing or timing constraints, or other "ilities"—maintainability, testability, portability, or other similar requirements that may impose constraints on the ontology design.

This summary is critical for semantically enabled systems development—we cannot emphasize enough how important it is to understand the context for ontology work. Developing the summary will be iterative; you'll add more content as you discover it through the analysis process, but documenting what you know from the start provides some grounding and can serve as the basis for conversations with subject matter experts. As we mentioned already, if the analysis is done in the context of a larger team, or as a part of developing a business or software requirements document, then some of the summary elements listed above will belong in other places in that document rather than in the use case itself. There will still be a need to document the scope of the individual use case in the context of the larger application, pre- and post-conditions for the use case, etc., however.

Common actors include end-users, data providers and information resources, other systems or services, and the ontology or ontology-enabled application itself. A primary actor typically initiates and is the main consumer of the service or information the use case provides. Secondary actors respond to the requests made, provide data, perform transformations on request, perform reasoning or analytics tasks, and so forth.

Preconditions define the conditions that must be true (i.e., describe the state of the system) for a trigger to meaningfully cause the initiation of the use case, or to initiate some subset of the process steps identified in the use case. Some level of modeling preconditions may be required as you develop your ontology, but typically not in a first pass encoding, which focuses on the main process flow, goal, description, etc. There are two cases to describe with respect to post-conditions: (1) the success scenario, where you should restate how the use case via its flows and actors and resources achieves the desired result, including any artifacts produced, impacts, and metric values; and (2) a failure scenario. In the failure case, it is really helpful if you can specify not only how the use case via its flows did not achieve the desired goal, but also describe the role of any actors and other resources in the failure, which artifacts were not produced, any known impact of failing, and associated metric values. Certain aspects related to failure may not apply in every case, but keep in mind that this list is intended to be as complete as possible, based on our experience.

3.5   THE "BODY" OF THE USE CASE

The body of the use case includes several sections, including a section that defines usage scenarios, process steps for the "normal" and alternative flows, use case and activity diagrams that flesh out the steps and interactions among actors involved in those flows, and a competency questions—the questions that the system using the ontology and/or knowledge base needs to be able to address. It is the latter that is unique to ontology engineering, but the concepts and relationships you model will be derived from all of these elements. Typically, we start with the summary and then identify a couple of key questions that the application, repository, or ontology must be able to answer, and build out from there.

Development of a good use case is anything but a one-person job, however. While one individual can provide a good initial draft, good use case design requires input from a variety of people. The next steps involve discussions with stakeholders—end users, project management, downstream recipients of any recommendations, reporting, or other intelligent/analysis products from the ontology-based system, information providers, and so on, to solicit user stories, get agreement on composite usage scenarios, and build an increasingly rich set of requirements for the ontology and any related application(s). Often, these initial steps are performed by others, typically business analysts or researchers driving the overall system development. However, a typical business analyst may not be aware of the need for collecting the reference documentation, models (i.e., business architecture and process models, any existing logical data models, etc.), competency questions, and information about participating resources that an ontologist needs. It's the role of the ontologist to actively participate in the requirements gathering process, to make sure that use case writers include the information needed to craft competency questions together with sample answers, to ensure good coverage. Additionally, you should make sure you understand and can document an illustrative process by which the answer is obtained so that you understand what kind of representation and reasoning is expected of the ontology and the ontology-powered application.

The results of this analysis, documented in the use case, should include:

- usage scenarios—at least two narrative scenarios that describe how one of the main actors would experience the use case, derived from some representative set of user stories;

- normal or primary flow process steps, including which actors, applications/ontologies, resources, and stakeholders participate;

- alternative process flows, such as initialization/set-up, cases where certain resources are not available, flows initiated by various events or sensor inputs, etc.;

- use case and activity diagrams—including a top-level or overview diagram and additional diagrams decomposing the high-level capabilities into subordinate capabilities (at least one level down, sufficient to identify additional actors and resources);

- competency questions—questions the knowledge base, application, or ontology should be able to answer in fulfillment of the use case *and a representative set of sample answers*; and

- a description of how the answer(s) to the competency questions may be obtained, including any reference to use of the ontology to answer the question(s).

Use case and activity diagrams are frequently built in Unified Modeling Language (UML), but they could be created in any tool your team is comfortable with (e.g., Visio, PowerPoint). We recommend use of UML, since the tools typically make it easy for one to create a far more complete model that can be used for downstream ontology and software development. We do not cover the diagramming process and tooling here, however, as the UML specification (2017), Jacobson et al., (1992), Jacobson, Spence, and Bittner (2011), and others provide excellent references.

3.6 CREATING USAGE SCENARIOS

Although we do not necessarily recommend Agile (Beck, 2001) and/or Scrum[60]-based development approaches for ontology development per se, an agile approach to soliciting requirements from stakeholders and SMEs through collecting user stories and developing narratives (usage scenarios) can be very helpful in requirements analysis for ontology development. The stories will provide at least some of the vocabulary required, and will assist in identifying boundary conditions, i.e., scoping requirements. They can also be used to provide the basis for reporting progress with respect to where you are in the development process, such as stating that the ontology covers x% of the terminology from the total number of user stories at some point in time.

A user story is a short, one or two sentence statement about some capability or function that a user wants to be able to perform, from a user's perspective.[61] User stories provide a catalyst for discussion about the behavior of a system that the user needs. They form the starting point for requirements analysis and decomposition. Any ontology-driven system will have many different kinds of users, with different perspectives and goals for the resulting system, so we need to engage at least a representative community of stakeholders in developing them.

We have found that a series of brainstorming sessions with subject matter experts and stakeholders is a good way to gather user stories. In some cases, starting with a more abstract, or coarser grained epic that we can break down into more detailed user stories, helps in framing the conversation. The kind of decomposition we typically perform varies somewhat from the functional decomposition approaches that a software developer would do, however. For example, suppose a user provides the following user story.

I would like to understand the numbers in this report and their implications.

Both an ontologist and a software engineer would want to know (1) what data is involved in the calculations, and (2) what formulas are used for calculation purposes. But, the ontologist may also need to know things such as the following.

What raw sources were used, how authoritative are they, were those sources updated recently enough for the problem being addressed, and are they considered the "gold source" for this information?

What systems touched this information before it entered this calculation process?

How are each of the variables in the formula defined, and where did those definitions come from?

What is a threshold that determines a "low (or high)" value, and what does it mean if a value is low (or high)?

Do the algorithms involved in the calculations have names, definitions, participate in any documented processes, or are they referenced in any policy?

What happens to this information once it is reported? Can we uncover any constraints that should be applied to ensure that the resulting numbers meet reporting requirements? What kinds of annotations are needed to support data lineage and governance requirements?

Usage scenarios are narratives that stitch a number of user stories together to describe how a particular end-user might interact with the system. They should be developed using a subset of the

---

[61] http://www.mountaingoatsoftware.com/agile/user-stories.

user stories that have something in common, involving the same actors and interfaces, with broader coverage than any individual user stories. This longer narrative is distinct from an epic, which is very terse, like a user story, but higher level. The goal is to develop a series of usage scenarios that incorporate the user stories you collect, and then review the usage scenarios with the stakeholders to confirm understanding and coverage. The scenarios will help frame the set of steps that together form the normal flow of operations for that use case, as described in Section 3.7. Based on our experience most use cases will include at least two usage scenarios, but typically not more than five, in order to manage scope.

Once you have a representative set of use cases summarized with a representative set of usage scenarios they should be reviewed with your stakeholders and against any requirements documentation to identify gaps, solicit additional user stories, and for prioritization purposes. Subsequent work on the use cases and on the ontologies they entail should be gated by project priorities at every stage in the development process.

3.7   FLOW OF EVENTS

Often referred to as the primary scenario or course of events, the *basic* or *normal* flow defines the process that would be followed if a system that implemented it (and participating actors) executed its main "plot" from start to end. Alternative processes, such as for first time start-up and initialization, states triggered by asynchronous events such as sensor inputs, or other states or error situations that might occur as a matter of course in fulfilling the use case, should be included as separate flows (Alternate Flow of Events in our template). A summary paragraph should be included that provides such an overview (which can include lists, conversational analysis that captures stakeholder interview information, etc.), followed by more detail expressed via a table that outlines the major steps and actors that participate in them.

The basic flow should provide any reviewer a quick overview of how an implementation is intended to work from a particular user's perspective. If there are multiple user perspectives, as is often the case with a large system, we would split the use case up so that each use case focuses on one primary actor's viewpoint. Determining what a given use case should include ties back to the user stories and narrative usage scenarios, such that each use case addresses only a single user perspective. The collection of use cases should provide coverage for all of the known primary actors.

Given that the use case is intended to provide the scoping requirements for ontology development, it is expected that all flows will include at least one step that includes consulting the ontology. Make sure to include a sentence about how the ontology will be used and give at least one concrete example of what information is obtained using the ontology in these steps.

In cases where the user scenarios, even for the same primary actor, are sufficiently different from one another, it may be helpful to describe the flow for each scenario independently, and then merge them together in a composite flow. Each element in the flow usually denotes a distinct stage in what will need to be implemented. The "normal" flow should cover what is typically expected to happen when the use case is executed. Steps are often separated by actor, intervention, or represent modular parts of the flow—a given step can encapsulate a set of activities that may be decomposed further through the analysis process. Use cases can incorporate iteration (loops)—often a group of steps will be repeated, so identifying each step will provide the ability to reference an earlier step in the flow. The normal flow "ends" when the final goal of the use cases is achieved. Consider incorporating one or more activity or sequence diagrams (or state diagrams) as a means to turn the written process flow into a visual one for stakeholder presentation, review, and validation. Any artifact, repository, or service that is needed to execute the flow should be documented in a resources section of your use case.

Alternate flows include variations on the basic flow, often invoked by valid, but not typical, situations or rules. These should include known "special cases," initialization, sensor-triggered flows, other anticipated alternate paths, and certain anticipated exceptions, such as cases where resources are unavailable to complete a normal processing flow, especially if there are back-up or alternative resources that can be used as replacements. If alternate ontologies or subsets of the ontology are required during the alternative flow processing, include a statement about what that portion of the ontology will be expected to help answer. Exhaustive coverage of error conditions is not required though. Activity diagrams are useful in representing these alternate flows case as well. Diagramming in this case can assist in identifying more general patterns that might well belong in the normal flow of this or another use case that involves other primary actors.

3.8   COMPETENCY QUESTIONS

The term "competency question" has been used for many years in the knowledge representation community to refer to a set of questions and related answers that a knowledge base or application must be able to answer correctly (Uschold and Grüninger, 1996; Grüninger and Fox, 1995). The idea is to ensure that the resulting system has sufficient background knowledge to successfully fulfill its mission. The term is derived from a similar notion in human resources, whereby an interviewer is interested in ensuring that candidates for certain positions have the requisite background for the job. The interviewer will have some pre-designed set of questions that they know the answers to, to test a candidate's competency to perform the job, including some that are considered trickier than others to answer.

Competency questions can be derived from the user stories, usage scenarios, normal and alternative process steps, as well as from reference materials. They should cover at least the primary, representative anticipated content for the normal flow, using the domain and stakeholder terminology. In situations where multiple sources of information are required to get to an answer, it is important to craft competency questions that span those sources for validation and regression testing purposes. We consider these questions and related answers to be even more important than the diagrams or alternate flows in terms of developing and validating a preliminary ontology, because we use them to help drive not only the scope of the ontology but the ontology architecture. Further, it is important to provide not only the anticipated answer for the question, but some short description of how that answer should be obtained. This description helps determine what information must be encoded in the ontology and in any application processing components. The questions and anticipated answers are also critical to ensuring that changes to the ontology or related system have not invalidated the content of the knowledge base, especially for long-term maintenance purposes.

For example, from the garden-oriented user story interested in oak-root fungus susceptible trees, competency questions might include the following.

> *Question: From the set of approved trees for the city of Los Altos, identify all those that are oak root fungus resistant, that flower, that are deciduous, and whose leaves turn colors in the fall.*

> *Answer: koelreuteria paniculata (golden rain tree), melaleuca stypheloides (prickly paperbark), lagerstroemia indicus (crepe myrtle), malus "Robinson" (Robinson crabapple), sapium sebiferum (Chinese tallow tree) … maybe others.*

> *Answer: Definitely not a cinnamomum camphora (camphor tree).*

Determining an answer entails: (1) downloading the list of trees;[62] (2) analysis to determine which of the trees on the list are oak-root fungus[63] resistant (requires confirmation through multiple garden guide sites such as those from local nurseries[64]); (3) which are deciduous (not hard—they are classified that way in the list); (4) which are flowering (may require a separate reference, such as the Sunset Western Garden guide, although some are identified in the list); and (5) which have showy fall foliage (again may require a separate reference). If the user provides additional information, such as where in the landscape the tree is intended for, some of those in the list might be eliminated if their roots are known to cause problems for nearby pavement or plumbing, for example. A recommendation system for planting options might also compare the results with information from other people living in that area who reported success (or failure) with replacing dead trees with some of those on the list, as well as from landscape architects and arborists in the area that would act as SMEs.

An ontology that models garden plants in such a way as to answer this question would need to encode information about plants and trees used in landscape settings, and their botanical details.

---

[62] https://www.losaltosca.gov/sites/default/files/fileattachments/Building%20and%20Planning/page/432/streetreelist.pdf.

[63] http://www.ipm.ucdavis.edu/PMG/GARDEN/PLANTS/DISEASES/armillariartrot.html.

[64] http://www.yamagamisnursery.com/newguides/Miscellaneous/Local_Oak_Root_Fungus_Resistant.pdf.

It would also need to encode information about pests, including naturally occurring fungus, deer, etc., and remediation strategies to address them. In a more complete application, information about soil, light requirements, how large the plant is anticipated to be in a landscape setting, and other gardening-oriented information, such as pruning recommendations, would be appropriate. The list of references and resources gathered to address these questions should be included in your use case under references and/or additional resources (such as the repositories one can use to get answers to parts of the competency question). Further, it is worth noting the effort required to answer the questions. It could be that the work involves checking values and thresholds of property values (e.g., the amount of water needed for a particular plant, and availability thereof). If data integration across multiple sources is anticipated, then additional mapping and ontology encoding will be required to specify the relationships between the terms in the participating resources. It is also useful to note whether inference is required, such as to suggest whether or not the plant or tree could grow beyond the available space in the case of our example, or possibly incompatibilities between situations or events.

3.9   ADDITIONAL RESOURCES

The resources section of the use case is also very important for ontology engineering. It should include information about every non-human actor, including a description of the resource, the information or service it provides, details about how that information or service is invoked and any transport details (access mechanisms/APIs, policies, usage policies, license restrictions, and requirements, etc.), who owns it, and so forth. A detailed list of criteria for describing resources is provided in our use case template for reference purposes.

There are always miscellaneous notes including details discovered through the use case development process that don't seem to fit anywhere else. We simply collect them as notes at the very end of the use case, and then move them around as needed once we understand where they fit into the overall requirements. That way the information is not lost, and can be shared with other members of the development team throughout the engineering process.

Finally, we remind our partners (and ourselves) to revise their use cases with every step of the ontology engineering process. Use cases are living documents that reflect not only preliminary requirements but details learned through development, testing, deployment, maintenance, and evolution. Because we are developing ontologies that may be shared in a public Semantic Web environment, evolution may include incorporating requirements from new stakeholders that were not considered initially. Keeping use cases up to date and publishing them with other ontology documentation ensures that people who might want to reuse the ontology understand the requirements and constraints that framed the development process.

3.10 INTEGRATION WITH BUSINESS AND SOFTWARE REQUIREMENTS

Many engineering projects, and especially software and data engineering projects, fail due, at least in part, to poorly specified, underspecified, or lack of requirements and evaluation criteria. In a product development-oriented organization, requirements gathering and analysis may be the responsibility of marketing, systems engineers, product managers, and sometimes software and data architects. The more mature the organization is, the more likely it is that they have a team that is dedicated to requirements development and process management. For these organizations, adding ontology-specific requirements, (e.g., additional references for terminology and definitions, competency questions, identifying reusable external ontologies, potential mappings between terms or relationships between vocabularies, and information about resources) may be a natural extension of current processes. The additional ontology requirements should be incorporated into a larger business requirements document (BRD), which typically describes the "*what*"—the desired capability from a business perspective, and/or a systems requirements document (which often provide a *response* to the business requirements—the "*how*"), or other similar document based on best practices and context. We should note that our usage of the term "business" is in its broadest sense, i.e., in the context of business management, and is not limited to commercial settings. Business management, project management, governance, and related activities, performed by "business-focused" or functionally focused people are as central to government, not for profit, and research organizations as they are to commercial endeavors. Organizations whose primary "business" does not include software or data engineering and management practices, meaning most companies other than technology or engineering companies, may or may not have a strong requirements development capability. In the absence of a well-written requirements document, use cases work well for capturing what is needed in terms of domain knowledge reference material, scoping requirements, basic functional requirements, and success criteria.

A typical BRD includes:

- problem context, scope, and a summary of the capabilities needed;

- primary stakeholders, including potential users, funding sources, upstream information and service providers, downstream consumers of the resulting capability, including indirect consumers;

- success factors for the envisioned capability or target system;

- constraints imposed by the organization, regulators, or other stakeholders, systems, or the environment for which the target system or product will be designed;

- use cases, including usage scenarios that provide context for how the system/product/service is intended to be used;

- business process models and analyses, with diagrams that explain "as-is" and "to-be" business processes;

- vocabulary resources (including glossaries, nomenclatures, data dictionaries, and the like) representing terminology pertinent to the capability;

- information flow diagrams to show how critical information/data flows through the required capability(ies) and existing systems;

- conceptual or possibly logical data models including any reference data models that provide at least the minimal information known about data requirements for the resulting capability; and

- wireframe diagrams that describe envisioned user interfaces (optionally, and depending on the context for the BRD).

A preliminary ontology could be incorporated into a BRD to provide a basic vocabulary, related definitions and the seed for elaboration not only of the ontology itself but for other project components, such as for logical data and business process models. Incorporating an ontology could happen in a second or third pass on the BRD, once at least some context is available to use as the basis for development of that ontology. Or, the ontology could be derived from the requirements in the BRD, but in that case one would hope that any logical data modeling would also be derived later and validated with the ontology for proper terminology use.

In the context of a BRD, use cases are typically a subset of the document. Information about the project, including overarching scoping requirements, the top-level architecture and interfaces, actors and stakeholders, and other high-level summary information are documented in the broader BRD, while only those details specific to an individual use case would be captured in the use case itself.

There is some controversy around how to develop use cases and in what level of detail among long-time practitioners of software and systems engineering. Some software engineers think of use cases as a tedious, often throw-away part of the analysis process, and many data architects want to go straight to modeling instead of thinking through a representative number of use cases. This seems to be especially true in cases where replacing an existing system is one of the primary reasons for the project. Yet, using an existing system as the primary blueprint for a new one often misses critical requirements of end-users and stakeholders (e.g., some of the reasons for replacement in the first place). In other words, skipping the analysis phase of the project can be a precursor to failure.

Even in the context of a more formal environment that relies on system or software requirements documents as the basis for service level agreements between the development team and project stakeholders, the content and level of detail included in the use cases varies. A typical software requirements document (SRD), derived from a system requirements document might include:

- an introduction that includes the problem statement, system overview, and scope;

- a system description, that includes any relevant interfaces (e.g., systems, users, hardware, other software or repositories), operational environment and constraints;

- a concept of operations, that provides a functional overview—i.e., describes how the intended system should work;

- detailed requirements by capability, including interfaces for specific functions within the overall product/environment;

- supporting use cases, including user stories, use case diagrams, activity and sequence/interaction diagrams;

- logical data models that specify the structure of the data required for use by the resulting software system; and

- traceability matrices that provide provenance to the original requirements.

Ontology-specific analysis and requirements development overlaps with both the "business-oriented" and software engineering approaches. We take a heavier-weight, rather than lighter-weight approach to use case development here, but it's primarily because we may not have the context of either a business requirements or systems requirements document to fall back on. Clearly in the context of a business analysis team effort, the project level information should be documented in the BRD, where it belongs. The ontology, at least the highest level concepts, terminology, and related definitions should be documented in the BRD as well, replacing the traditional glossary. And, in a software engineering context, the ontology requirements and use cases would be part of the broader SRD. No matter how the information is documented, however, we need to understand the "*why*" and the "*what*"—including any background knowledge and reference information, to construct an ontology that meets stakeholder needs. It's from the "business" or functional perspective that we learn:

- the problem context, scope, and desired capabilities;

- who the primary stakeholders are, especially for those using/consuming the resulting system;

- what it means to be successful;

- constraints on the vocabulary that will be incorporated into the ontology, particularly any required terminology for use in question answering, analysis, or reporting; and

- relevant vocabularies/glossaries that are used in the domain or by the organization.

We also depend on any use cases, including usage scenarios that provide context for how the system/product/service that implements the ontology (if it is implemented in the context of a system) is intended to be used. Business process, information flow, and data models that already exist, can be used to inform the ontology (both for terminology extraction and to provide context).

From software engineering, especially when developing ontologies for semantically enabled applications, we also need:

- the context in terms of *interfaces* to users, systems, software, and information, so that the ontology delivers any required user or system facing vocabulary as well as mediation capabilities across siloed information;

- the context in terms of a functional overview—so that the ontology may be designed to support the required classification and question answering capabilities needed in context to support those functions;

- detailed supporting use cases, including user stories, use case diagrams, activity and sequence/interaction diagrams, that arise from preliminary software architecture and requirements analysis activities that may provide additional terminology or other requirements for the ontology; and

- any additional models that specify the structure of data that may be used by the resulting system to answer questions, make recommendations, perform analytics, etc.

In addition to all of the above, we can also extract terminology and constraints (restrictions and other logical axioms in an ontology language such as OWL), from other resources. These include:

- reference documents—policies, procedures, business documentation, standards, terminologies and glossaries; and

- any known "business rules," algorithms, or sources of statements about how the information is structured and used.

Because we are less concerned with the platform and infrastructure on which the ontology will be deployed, certain non-functional and hardware related requirements are typically less important to an ontologist. Having an understanding of such constraints never hurts, though. Certain kinds of logical axioms are more costly than others from an inference and query performance perspective, and if results need to be provided within a certain window of time, we need to know that.

CHAPTER 4

Terminology

In this chapter, we introduce terminologies and their management, for example, as in (Wright and Budin, 2001; Ji, Wang, and Carlson, 2014), which are essential components of ontology engineering.

Terminology work is concerned with the systematic collection, description, processing, and presentation of concepts and their designations (ISO 1087-1, 2000). By terminology we mean the study of terms, which may be individual words or multi-word expressions, that stand for concepts within a specific context. There are many definitions for *concept*, but we use the ISO definition, which is a "unit of knowledge created by a unique combination of characteristics." A terminology is a set of designations used together in a context, which may be a business context, a discipline, or other community, and which may be expressed in a single natural language or multiple natural languages. A designation, as defined by ISO, is a representation of a concept by a sign that denotes it, where a sign can be a term or symbol. So, terminology work involves concepts, terms that are used to refer to those concepts in specific contexts, and formal vocabularies, including the establishing definitions for those concepts. From our perspective, it also involves providing the *pedigree* (i.e., the history or lineage) and *provenance* (i.e., the source(s) and chain of custody) for the concepts, relationships, and definitions specified in an ontology. Understanding the pedigree and provenance for a piece of data, or for an artifact or document, provides the basis for transparency and trust.

We're often asked why we believe so strongly that a formal approach to the terminological aspects of ontology engineering is important. After all, for many common applications of semantic technologies, people don't actually see the terms and definitions in the ontologies or vocabularies; typically, they see only the results of question answering or recommendation services. Getting the relationships and logic right is essential, but the terminology? Virtually every ontology *we* have ever built and most that we are familiar with were designed to be used by people *and* machines. In some cases, such as ontologies for data governance, lineage, and management purposes, and particularly to support searching and natural language processing (NLP), designing the terminology to seem natural to a wide range of users may be more important than fleshing out detailed logic axioms. Even in cases where the usage scenarios for an ontology are computation oriented, such as for recommendation systems, machine learning, predictive analytics, and so forth, the terms and definitions must be *accurate*, *precise*, well-supported via *pedigree* and *provenance*, and make sense to qualified SMEs. Many technologists make the mistake of assuming they know enough about a business domain to encode the language used in their ontologies or other models (e.g., business process models, information flow diagrams, schema, software architecture models, state diagrams) without the assistance of SMEs. We do not. We firmly believe that the terminology used in an ontology must reflect the language of the organization and domain, as validated by the appropriate expert(s).

As an example of a case where terminology is essential, consider the financial crisis of 2008 and its aftermath. Reports by banking regulators suggest that the weakest links in banking compliance data are due to problems in data architecture and information technology (IT) infrastructure.[65] Among the action steps recommended by the Basel Committee on Banking Supervision (BCBS, one of the primary regulatory authorities for the financial services industry in the European Union) are "identifying and defining data owners" and "consistent and integrated [business] [controlled] vocabularies at the group and organizational level." The root cause for these issues is typically not technology per se, but terminology. In many financial institutions, common vocabularies are either inadequate or non-existent. As a consequence, there is no *ground truth* for many banking data assertions or the decisions that depend on the data (i.e., no basis for agreement based on empirical observation, provenance, and pedigree), since the underlying, common semantics are missing. There is no concurrence within groups or across the organization, let alone across trading partners, regarding the meaning of the terms used to assess risk. And, there is nothing to guide IT departments in these institutions in the construction of data-oriented applications and services because the business requirements are vague and often circular. In most of these cases there is no methodology in place to address the terminology vacuum.

- It is deferred to IT when in fact it is a business problem.

- There are no resources or expertise in-house to address it.

- SMEs are otherwise engaged (e.g., in the business of trading, where the people who understand the business they are conducting best have very short attention spans and even less time to assist).

A shared understanding of the meaning of the terms is what makes compliance, and many other capabilities, tractable. For example, the BCBS statement on "Principles for effective risk data aggregation and risk reporting,"[66] also known as BCBS 239, requires "integrated data taxonomies and architecture across the banking group," i.e., a ground truth for risk data. In this context, risk data aggregation is defined as "defining, gathering and processing risk data according to the organization's risk reporting requirements to enable that organization to measure its performance against its risk tolerance/appetite." BCBS 239 mandates internal standardization of metadata, institution of single identifiers and/or unified naming conventions for concepts and related data elements (e.g.,

---

[65] Basel Committee on Banking Supervision: Progress in adopting the principles for effective risk data aggregation and risk reporting, January 2015. See https://www.bis.org/bcbs/publ/d308.htm, and the follow-up reports released in October 2016, available at https://www.bis.org/bcbs/publ/d388.htm and April 2017, available at https://www.bis.org/bcbs/publ/d404.htm.

[66] https://www.bis.org/publ/bcbs239.pdf.

legal entities, counterparties, customers, accounts), and "adequate controls throughout the lifecycle of the data and for all aspects of the technology infrastructure."

A common vocabulary provides the basis for shared understanding and ground truth. It should be comprised of a unified, logical, and consistent approach to concept identification, naming and specification, through a combination of formal definitions, relationships, and additional metadata (e.g., explanatory notes, examples, pedigree, and provenance). A well-formed business vocabulary provides the ability to reconcile and integrate information across disparate systems and data sets. The pedigree, provenance, and definitional information enable organizations to exercise data governance, support data lineage analysis, and tie dynamically changing data back to requirements. Representing a business vocabulary as an ontology enables automation of these requirements—permitting use by people *and* machines, for conversation and communications as well as search, question answering, integration, business intelligence and decision management, recommendation services, predictive analytics, and other research purposes.

4.1   HOW TERMINOLOGY WORK FITS INTO ONTOLOGY ENGINEERING

Development of a vocabulary for some domain may include several interdependent but parallelizable activities that can be phased by sub-domain, user stories, or by addressing specific competency questions. Generally, the steps involved in ontology development, including terminology work, are roughly as follows.

1. A preparatory phase, involving: identification of a specific sub-domain, topic area, or organizational perspective to focus on, gathering relevant documents, including any use cases if they exist, and other references, such as controlled vocabularies, relevant glossaries, related policy documents, etc., and identifying key stakeholders, principal SMEs, and consumers of the work product.

2. A term excerption phase, which involves extracting keywords and key phrases from the content provided in phase 1 to create a preliminary term list, with preliminary definitions and other annotations, including source and context annotations for every term. We apply automation to this process to the degree possible, using extraction tools that leverage NLP methods.

3. An optional business architecture development phase, covering the primary value streams and value stream stages, a breakdown of capabilities corresponding to those value streams, and identifying the information entities that participate in each capability, with examples, sources, requirements, and any dependencies among them.

4. A subsequent term excerption and normalization phase, which entails extracting keywords and key phrases from the business architecture, including the interviews conducted to augment the preliminary term list, if a business architecture is available. Business process, enterprise architecture, and information models are also useful as inputs to the normalization process, and terms should be cross-referenced to any such models where appropriate.

5. A use case development phase, including usage scenarios and competency questions (meaning, the set of questions that the use case entails—questions that a repository needs to be able to answer, an application needs to cover, example answers to those questions, some description of how to determine answers, etc.). If a business architecture is available, then the use cases must each carve a path through the capability and information maps, such that every capability and every information element is addressed in at least one use case.

6. A term list curation, reconciliation, and augmentation phase, which involves identifying the subset of terms in the term list and derived from competency questions that are considered most critical by SMEs and stakeholders. The terms should be prioritized, and the highest priority terms should be used as the basis of expansion. This phase also involves mapping the terms to concepts, potentially in external ontologies that have been identified for reuse, definition reconciliation and refinement, and capturing metrics to quantify insights into the nature and number of changes made during curation to assist in predicting ontology development and evolution costs. Note, too, that any term that falls out of analysis of the question answering process must be included in addition to the basic terms that appear in the questions themselves.

7. An ontology development phase, taking the business architecture, use cases and term list as input and creating a formal, conceptual model that contains the concepts, definitions, other relevant metadata, relationships among concepts, including relationships to externally defined concepts from ontologies determined to be reusable, and formal axioms that distinguish the concepts from one another.

8. A follow-up review phase, in which the SMEs and stakeholders review the resulting ontology for correctness and completeness, including validation with test data that can be used for regression testing as the ontology evolves.

9. A deployment phase, in which the ontology is made available via the organization's intranet (or external web) for broad access, and mapped to back-end repositories and data stores, and/or integrated with applications, based on the use cases.

The process we teach our students excludes steps 3 and 4, above, but covers the remainder, which corresponds to what we cover in the rest of this chapter and in the book as a whole. In addition to these steps, ontology engineering work doesn't stop at deployment, but includes maintenance and evolution, which are extremely important for fostering living user communities around any ontology. In cases where the ontology may have a long life and broad usage, we also recommend creating and documenting policies and procedures for developing, managing, evolving, and governing the products from each of these steps. Depending on the project requirements, we also recommend documenting policies and procedures for *using* the ontology in data governance, IT requirements, regulatory compliance, downstream applications, and for other purposes.

As we've said before, the business architecture, if available, and use cases are crucial to determining the scope of the ontology. They identify key stakeholders, information entities, and provide a baseline set of terms as input to the term list development activity. The use case documentation must cite all reference materials—regulatory and internal policy documents, international and national standards, other relevant references such as dictionaries and other documents for definition development—source, and product repositories, and artifacts that are relevant for that use case. These references fill in some of the gaps that a business architecture, other enterprise models, or other requirements might not naturally cover. The use cases must also include the competency questions (questions that must be answered to fulfill the goals of the use case, together with a sample of expected answers and a description of how those answers can be derived, if not obvious). The competency questions and answers, together with SME prioritization of the terms developed in step 6 help us identify the concepts and relationships required, understand how much detail is needed in the ontology, and most importantly when to stop.

Deliverables out of the methodology outlined above include:

- the use cases including relevant competency questions;

- the curated term list, including definitions and other annotations;

- ontology artifacts, including models the OWL, diagrams (which we recommend creating in the UML), test cases and test individuals, and documentation;

- the business architecture deliverables (value streams, capability maps, information maps, interview questions and responses, gap analysis, and so forth), as appropriate; and

- related policies and procedures, including some sort of statement to the potential users of the ontology regarding what to expect from a maintenance and evolution perspective.

Figure 4.1 provides a high-level diagram outlining the overall ontology development process in a typical business setting, including inputs, outputs, stakeholders, and business architecture, with a focus on terminology.

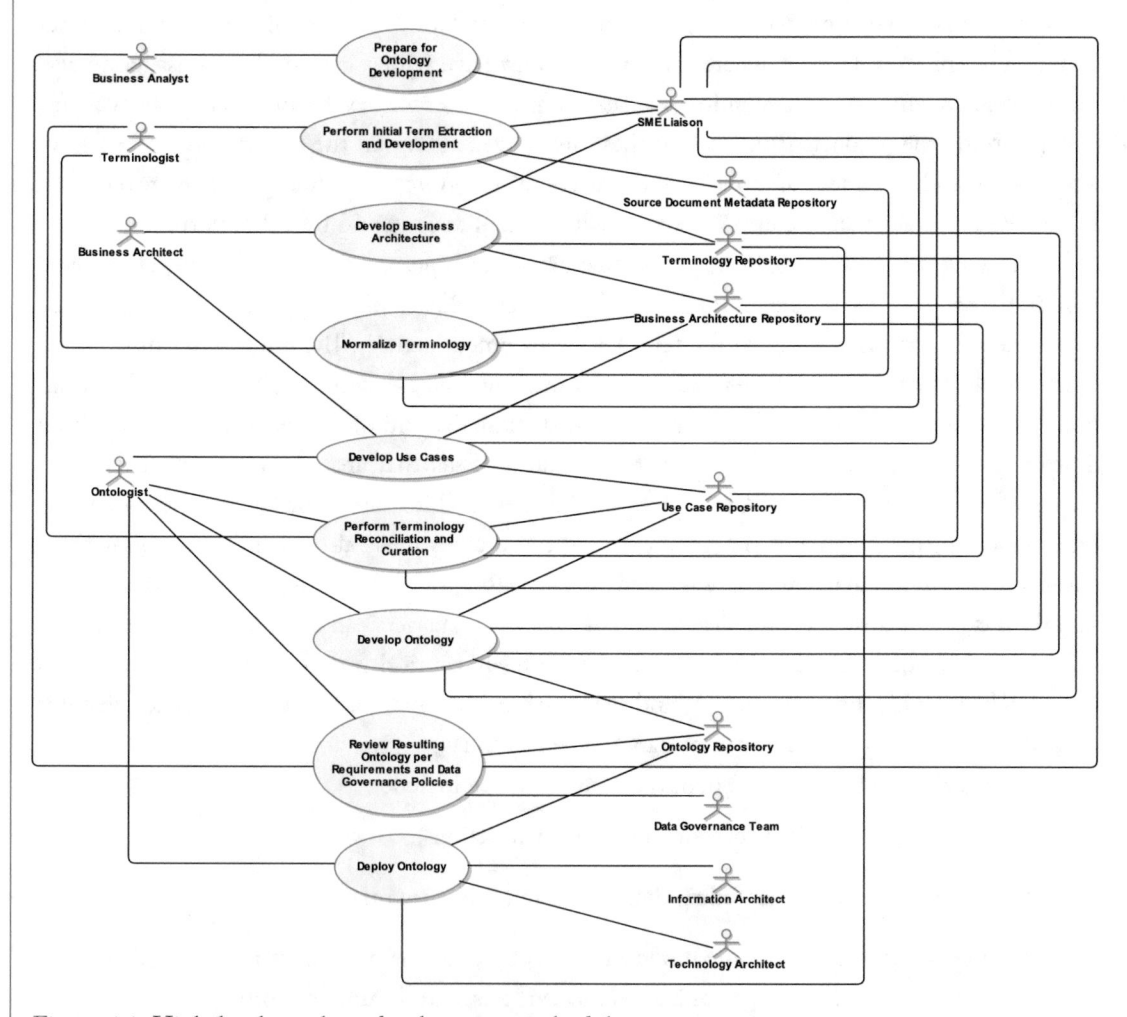

Figure 4.1: High-level ontology development methodology.

What the figure does not show is that some of the work can be done in parallel, such as term excerption and use case development. In larger teams, and especially in cases where there are many policies, procedures, and other documents to evaluate, doing at least some of the work to extract terms in parallel can save time. How much can be done in parallel necessarily depends on the skills of the team. The skills of some of the actors identified in the diagram, such as those of terminologist, ontologist, and information architect, overlap, and frequently some of the related activities are accomplished by a single person, but identifying the individual roles and accompanying skills can be helpful for planning purposes.

4.2   LAYING THE GROUNDWORK

Typically, an organization has completed at least part of what we call the preparatory phase prior to deciding that they need an ontology, but not always. Sometimes we get involved early to help people identify the things that they need to gather to move things along. We have found that scoping an ontology development project can be a daunting task for some people who have not done it before. Development of a preliminary set of use cases at a high level—including developing a preliminary set of questions that need to be answered using the resulting knowledge base and data sets is often a pragmatic, approachable place to start. In other cases, we've started from a business architecture perspective, identifying the capabilities that an organization has and then using the terminology derived from the capability and information maps as an entry point. Regardless, at the end of the preparatory phase we need to have:

- a reasonably clear idea of the scope of the domain area to be addressed, goals of the project and expectations for the resulting ontology, including some understanding of the scope of the questions that the ontology will be used to answer;

- a preliminary set of documents and artifacts that can be used as input to the process, including as many existing vocabularies, process and procedure documents and models, documentary artifacts and screens, as well as any known standards and best practices; and

- a list of the key subject matter experts and stakeholders for the project, including their roles and areas of expertise.

The collection of documents should include

- any standard controlled vocabularies and nomenclatures available for the domain, prioritizing those that are particularly important to use;

- any standards-based, or well-known hierarchical and or taxonomic resources that are relevant;

- any standard dictionaries for the domain, such as Barron's for finance and business terminology;

- reference documents including any relevant international (e.g., ISO, IEEE, OMG, W3C, OASIS, etc.) or national (e.g., ANSI, US government agency glossaries, regulations, policies, and procedures for the domain) standard and de-facto standards; and

- best practices documentation that are either descriptive or normative in nature.

For the landscape examples described earlier, controlled vocabularies might include the Royal Horticultural Society Color Chart. Taxonomies for gardening include the Linnaean taxonomy and of course the Plant List—from the Royal Botanic Gardens, Kew, and Missouri Botanical Garden. There is also a very large knowledge graph (thesaurus) under development by the National Agriculture Library at the USDA that would be appropriate as input,[67] in addition to more general ontologies for botany and biology found in the BioPortal.[68] Other reference documents for those of us who live on the west coast include the Western Garden Book[69] and related publications from Sunset Magazine, as well as research from the University of California at Davis.[70] In fact, it is information from UC Davis that enables us to answer questions such as:

> *What deciduous trees native to northern California are drought tolerant, resistant to oak root fungus, and grow no taller than 20–25 feet?*

4.3   TERM EXCERPTION AND DEVELOPMENT

The term excerption phase involves identifying keywords and key phrases from the materials gathered during the preparatory phase. These may include vocabularies, glossaries, policies, procedures, process and business architecture artifacts, standards, best practices, and other documentation known to subject matter experts, which should be the basis for creating a preliminary term list, with preliminary definitions and other annotations. As we have already mentioned, maintaining the pedigree and provenance for every term, including the source of the definitions, in the form of annotations, is essential to the ontology development process.

We recommend that certain information (which could be in the form of spreadsheets, a database, or knowledge graph) be maintained throughout the excerption process, including:

1. source document metadata, describing the document and providing bibliographic details, including URIs if it is available online (and for online documents, when they were retrieved), as well as who provided the documents and when, where they are maintained in an archive, and so forth; and

---

[67] https://agclass.nal.usda.gov/dne/search_sc.shtml.
[68] https://bioportal.bioontology.org/.
[69] https://www.sunset.com/garden/new-sunset-western-garden-book.
[70] https://caes.ucdavis.edu/research/centers/ccuh.

2. a term list that for every term includes the term itself, candidate definitions, annotations about the term, the context and community in which it is used, the source (and where in the source) it was retrieved and/or derived from, and so forth, and process related annotations that provide a view on the stage in the review process for a given term, at a minimum.

To assist our students in getting started, we provide a template for collecting terms, definitions and related annotations.[71] We also apply automation to this process to the degree possible. We have used a variety of NLP and text analytics tools to extract terms from documents to jump start term excerption. The results of automation are rarely perfect, and require human curation, but can provide significant benefits in terms of time savings. Figure 4-2 provides a view of the basic capabilities involved in term excerption.

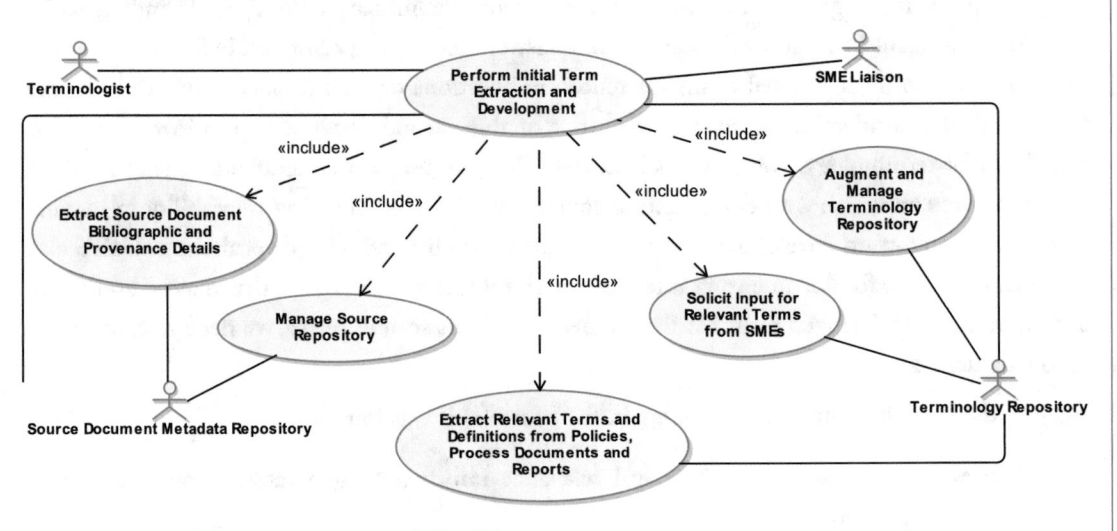

Figure 4.2: Initial term excerption and development.

Text processing tools that use NLP and information extraction algorithms to extract terminology from documents can identify the most relevant terms across a corpus of documents. They typically provide the terms themselves, their location within the document, and often a statistical notion of relevance to some subject area provided at the start of the extraction. Some engines allow the user to provide a ceiling on the number of terms, and can output results in a comma delimited list, spreadsheet, or other form. We typically limit the results to those terms that have at least a 50% likelihood of being relevant in an initial pass, and to a maximum of 100 terms per document. For a corpus of five to ten documents, that works well, but for a large corpus, more restrictive limits may be needed.

[71] Our terminology template can be downloaded from the Morgan & Claypool book abstract page, here http://bit.ly/2IpFOGp.

For an initial pass at a "raw" term list (meaning, a list that has not been curated in any way by subject matter experts), selecting the most relevant 150–200 terms from the results of the automated extraction process provides a good starting point for further work. The terms should be managed in a term list (or repository of some sort) that includes not only the terms themselves, but information about source from which they were extracted, including *where* in the source they were found as well as the relevance indication. The details with respect to the location in the document can be very helpful when analyzing the context for usage and for the purposes of definition development. Duplicates, plurals, and other forms of the same term should be eliminated from the raw term list, and the terms should then be prioritized with subject matter experts and to confirm relevance.

The next phase of the work involves taking the "raw" or basic term list and working with SMEs to prioritize, augment, and validate the terms and preliminary definitions. If limited access to SMEs is available, a review of questions and answers should be performed before reaching out to SMEs to identify additional terms referenced in questions or used to answer questions before this round of prioritization is done. The output of this second phase is a prioritized term list. Standards for terminology work such as ISO 704 (2009) and other methodologies such as IDEF5 (1994) provide approaches to documenting source materials and ensuring traceability in moving from extraction toward development of a final curated term list and related vocabulary. IDEF5 also includes a template for documenting interviews with subject matter experts that may be useful. For each term in the prioritized term list that comes out of this second phase, we need to capture the following details:

- the term, its source(s), and where in the source(s) it was found;

- relevance, including any statistical relevance information generated through automated extraction;

- priority from a SME perspective;

- any preliminary definition(s), including any references for those definitions and any indication of preference for one definition over others provided by the SME(s);

- any information we can gather about the domain and scope of usage;

- example questions and answers using the term, tagging the terms with an identifier for the questions in which they were used as appropriate;

- key stakeholders, ownership, maintenance, resources for data sources that use the term; and

• any additional information that the SME can provide—places to look for additional resources related to the term, other SMEs to confirm details with, any notes about the term, and especially any critical standards or policies that define it.

The results of this prioritization and preliminary development process should be managed in some sort of terminology or metadata repository for reference. Again, we encourage our students to use the template cited above, but in a commercial or research setting we typically use a database, wiki or other online tool to facilitate collaboration with our SMEs.

4.4   TERMINOLOGY ANALYSIS AND CURATION

Once a prioritized term list is available, an additional round of analysis, curation, reconciliation, and augmentation is needed to identify terms that are involved in asking and answering the competency questions contained in the use cases, or that are considered critical by SMEs and stakeholders. This analysis effort integrates any additional terminology derived from the use cases that may be needed to explain results, or that provide background knowledge involved in the context of the questions, and assists in prioritization for modeling purposes. It often includes additional input from the SMEs to identify any further considerations needed to answer the questions. It also involves mapping terms to concepts, and developing and reconciling definitions for those concepts where more than one term/definition exists. Definition refinement requires SME participation, and we encourage sourcing definitions from standards and reference materials where possible. For one finance application for capital markets, we started with a single term, *trade*, and, after definition development, use case refinement, preliminary conceptual modeling, and ultimately ontology development, we ended up with roughly 300 concepts that were essential to defining what a trade *is* under various circumstances.

Metrics that provide insight into the nature and number of changes made during this analysis and curation process may be useful, so we recommend capturing additional details in the form of annotations. This includes tracking who made changes to the list (and to a specific concept), when, and why, and potentially who approved the definitions. These metrics can be used not only to quantify insights into the change management process, but to determine which concepts are the most stable from a publication perspective, and predict ontology development and evolution costs. A high-level diagram of the activities involved in terminology analysis, curation, and reconciliation is shown in Figure 4.3.

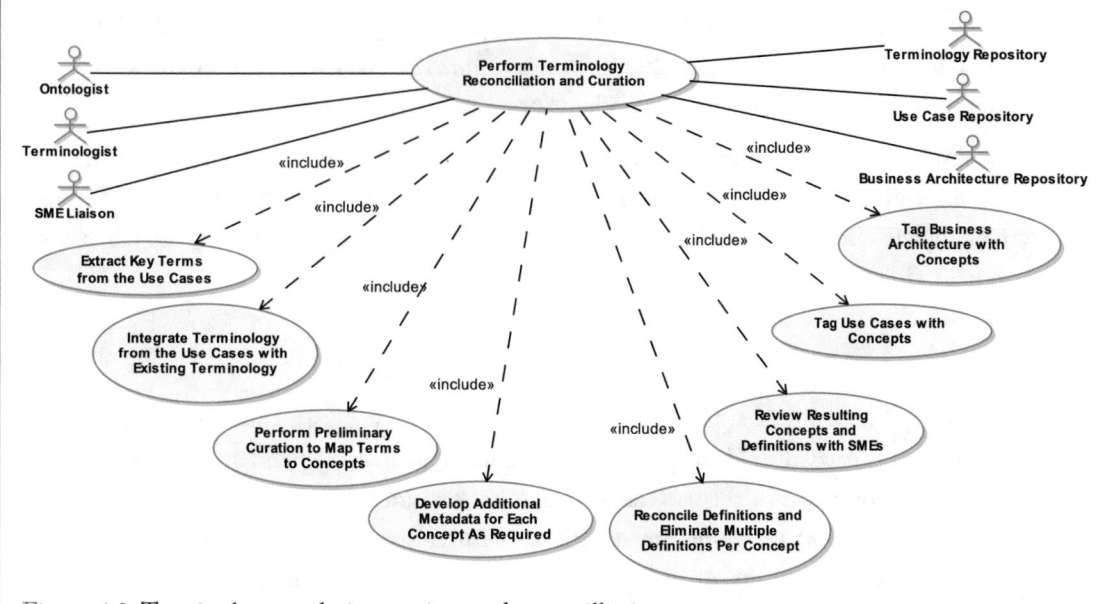

Figure 4.3: Terminology analysis, curation, and reconcilliation.

We follow the ISO 704 (2009) approach for defining terms and relating them to concepts, to the degree that we can. ISO 704 uses a model from ISO 1087 (2000) for capturing and documenting terminology and related vocabularies. In addition to providing guidance for taxonomy development, ISO 704 describes how to phrase definitions. It also makes recommendations for relating terms to one another using a standard vocabulary, which is documented in ISO 1087. ISO 1087 is a great resource for language to describe kinds of relationships, abbreviations and acronyms, related terms and designations, preferred vs. deprecated terms, and so forth. For definitions in particular, the standard suggests a "genus/differentia" structure, meaning, it recommends identifying one or more ancestral concepts as well as relationships and characteristics that differentiate the concept in question from sibling concepts.

- For concepts (nouns, which will become classes in OWL), most definitions should be phrased "a <parent class> that …", naming the parent(s) and including text that relates that concept to others through both relationships and characteristics (attributes from a data modeling perspective).

- For relationships (verbs, which will be properties in OWL), most definitions should be phrased "a <parent relationship> relation [between <domain> and <range>] that …"—in a similar form as the definitions for classes.

- Ideally, you should be able to replace a term with its definition in a sentence.

From ISO 1087, a concept is defined as a "*unit of knowledge created by a unique combination of characteristics.*" A vocabulary can be considered a system of concepts that are defined for a specific domain that are relevant to the organization. A vocabulary includes, for each concept, one or more commonly used labels, synonyms (and the context for their usage), formal text definitions, relationships with other terms, other constraints or relevant expressions, and additional metadata, such as, what report(s) they are used in, stakeholders for the report(s), relevant sources for definitions and for formulas for calculating the terms, where applicable. Typically, a *terminology* is defined as a subset of a vocabulary, limited to the nomenclature used by a smaller team or by natural language. In other words, although a vocabulary may be multilingual, a terminology is usually limited to a single natural language and conforms to the language used by an individual working group.

Concept names may be human readable or not, depending on organizational requirements and the development methodology used to create the vocabulary. In cases where the "same" concept has multiple meanings, some sort of mechanism (e.g., a superscript or suffix), should be used to differentiate them in preliminary term lists, at least until they can be reconciled or the distinctions can be addressed (often resulting in the creation of a separate concept per meaning). Duplication with overlapping or related definitions are common, especially in situations where the vocabulary is developed in a bottom-up fashion from data repositories or silos. Ultimately, however, the goal is to have one to one alignment between the set of concepts and definitions, without duplication.

A formal vocabulary is a "*set of concepts structured according to the relations among them.*" For our purposes, a fully curated vocabulary includes the following information for every concept:

- one or more commonly used labels (including abbreviations);

- synonyms (and the context for usage);

- formal text definitions;

- relationships with other terms, including relationships to externally defined terms;

- known characteristics (attributes);

- other constraints or relevant expressions, for example, formulas for calculating values, indicators of metrics, units of measure, as appropriate;

- explanatory notes that augment the definition;

- examples; and

- provenance and pedigree information including additional metadata, such as what report(s) the concept is used in, stakeholders for the report(s) as well as for the concept, relevant sources for the definitions as well as for any related formulas or calculations, who contributed content as well as who approved it, etc.

We will spend the remainder of this chapter elaborating best practices in fleshing out a vocabulary that includes the elements listed above as the basis for ontology development.

4.4.1   CONCEPT LABELING

Labels are typically specific to a community, which may be multilingual, but more often are unique to a smaller group of people within a large organization (for example, finance may use different labels for some terms from manufacturing or engineering teams). Jargon can arise even in small departments, and is often derived from the software applications that people use to perform their work. Several distinct labels may be used across a larger community for the same concept; the key is to determine if one or more labels are preferred over others, and to identify the context in which they are preferred.

Common metadata for labeling includes:

- a preferred label: the preferred human-readable label for a term; labels should be plain text, with proper spacing between words, and in lowercase unless they represent proper names or abbreviations (e.g., acronyms);

- alternate labels: when multiple labels are used to refer to the same concept, they should be recorded as alternates, but with metadata that describes where they are used (by community, document corpus, software systems, etc.);

- abbreviations: including clipped terms, initialisms, and acronyms are not separate concepts; if the abbreviation itself is the most common label for a concept, then it should also be the preferred label, with the fully expanded abbreviation in an alternate label;

- context: identification of the community that uses the concept with a given preferred label; and

- source: identification of any reference documents or sources for the labels.

4.4.2   DEFINITIONS

ISO 1087 defines the term "definition" as the "*representation of a concept by a descriptive statement which serves to differentiate it from related concepts.*" Reaching consensus within an organization for the definition of any concept can be more challenging than one might think. And, arriving at that consensus and documenting it is one of the most important activities in developing a common vocabulary. Definitions should be unique, not overly wordy, and not circular (i.e., concepts should not be defined using the concept itself within the definition). Definitions for compound terms should begin by referencing parent (more abstract) concepts. For example, the definition for *unilateral contract* from the FIBO[72], is "*a contract in which only one party makes an express promise, or undertakes a performance, without first securing a reciprocal agreement from the other party.*" Definitions should also incorporate any information regarding related terms or constraints, so in the case of a unilateral contract, the relationship between parties in intrinsic to the definition, and thus is important to include.

Definitions should also be terse, written in such a way that they can replace the term in a sentence. Any additional details, secondary explanatory information, examples, scoping information, etc. should be documented in other annotations about the term (i.e., notes).

Developing definitions isn't necessarily easy to do, and often the work required deferred, especially by teams that are more focused on the computational aspects of an ontology. That may be OK for prototyping activities, or for ontologies that are not intended for reuse, but in our experience, putting this off often means that it doesn't get done. The result is that users of the ontology are not clear on what any concept with a poor definition means. If one of the goals of developing an ontology is to facilitate clear and unambiguous communications, then an ontology with poorly expressed definitions fails to meet that goal.

4.4.3   SYNONYMS

Synonymous terms may or may not result in separate concept definitions, depending on requirements, their precise definitions, and the context for usage. For top-down vocabulary development efforts, the best practice is to capture synonyms as additional "labels" for a concept initially, and then as the definitions are formalized and use cases extended, a determination can be made as to whether the synonym should be included as a separate concept.

For each synonym, it is important to document:

- the synonymous term itself, in plain text as used, similar to a preferred or alternate label; and

- the community context: identification of the community that uses the synonym and for what purpose, if possible.

4.4.4   IDENTIFIERS AND IDENTIFICATION SCHEMES

In cases where concepts are used in existing systems, or are defined in formal vocabularies or ontologies already, there may be unique identifiers for them. The form that identifiers take varies widely, depending on the source systems or environments. Some vocabularies, such as those developed using Semantic Web languages, use Internationalized Resource Identifiers (IRIs), while others use

---

[72]   See https://www.omg.org/spec/EDMC-FIBO/FND/, Financial Industry Business Ontology: Foundations (FND), v1.2, Table 10.60 Contracts Details, p. 133.

unique identifiers that are language, environment, or application dependent. Many systems use structured identifiers that themselves may provide additional semantics. We call the specifications that define identifiers "identification schemes." Some common schemes in the U.S. include social security numbers, employer identification numbers (EINs, for corporate tax purposes), CUSIPs (Committee on Uniform Securities Identification Procedures) for identifying financial instruments such as stocks, bonds, commercial paper, and so forth, telephone numbers and Internet protocol (IP) addresses.

When identifiers for terms, definitions, synonyms, or other related information exist, it is important to capture and retain them, especially if the concept itself is retained as a part of the resulting ontology. Identifiers are particularly useful for mapping ontologies to one another to support interoperability, search, and other applications that involve multiple data sources. For every concept that has an existing identifier, it is essential to capture not only the identifier itself, but where it is used and any details that describe how it is defined, i.e., the related identification scheme.

4.4.5   CLASSIFIERS AND CLASSIFICATION SCHEMES

Some concepts may also be associated with classifiers in a classification scheme or codes in a coding system. For example, industry classification schemes, such as NAICS (the North American Industry Classification System), are widely used by statistical agencies for collecting, analyzing, and reporting on the status of the economy. Classification schemes, such as ISO 10962 CFI standard and ticker symbols, are used for classifying financial instruments. The Linnean taxonomy provides a scheme for organizing plants and animals, by kingdom, phylum, class, order, family, genus, and species. Some of these schemes have been expressed as ontologies that are publicly available, for example in the BioPortal. Even in cases where there is no public ontology, well-known classification schemes should be referenced when possible, as they can then be used to assist in integration and interoperability of the data sets that use them. The place to reference them initially is as a part of the curated terminology in annotations.

4.4.6   PEDIGREE AND PROVENANCE

Documenting the pedigree (i.e., the history or lineage) and provenance (i.e., the source(s) and chain of custody) for the concepts, relationships, other terms, labels, and especially for the definitions specified in a curated vocabulary can be as important, depending on requirements, as developing good definitions. Provenance associated with any terminology or ontology includes:

- the context or source for the concept itself;

- the context or source(s) for preferred and alternate labels;

- the context or source(s) for the definition, including any references from which the definition may have been adapted; and

- references, particularly to regulations and regulatory glossaries, corporate policies, standards, or other domain-specific papers and documents

Pedigree may include change history for the ontology itself as well as for any data sets that reflect that ontology including, but not limited to, what processes and systems have touched the data. The pedigree associated with a terminology or ontology, either at the ontology level, entity level, or at the data level may include:

- concept, term, or data element status: for example, proposed, provisional, approved, revised, deprecated, obsolete;

- status date: the date a concept, term or data element achieved a given status, i.e., was proposed, became provisional, was approved, etc.;

- modification date and by whom;

- steward: the name of the individual associated with the ontology or data set;

- change notes: any additional information associated with the changes that have been made; and

- approval information: in some cases, different authorities are identified to give sign off on a term definition for inclusion or update for the next release. Information may be captured concerning pending approval, or who approved the definition and when.

Depending on the size of the team, it may also be important to preserve and manage information related to the collaborative development process, including but not limited to review notes and discussion history.

4.4.7   ADDITIONAL NOTES (ANNOTATIONS)

In the process of developing definitions, often additional information is developed that doesn't necessarily fit as a part of the formal definition. This includes:

- notes on the scope of coverage of the concept;

- explanatory notes;

- notes on how the concept should be or might be used;

- examples; and

- notes describing any dependencies that the concept has on other concepts.

These notes can be very important for users to understand how to leverage the vocabulary or ontology, especially over time and as the body of usage notes grows. Documenting any additional references used in the process of developing the definitions may be useful to others, and so tracking the most important references is also good practice.

4.5   MAPPING TERMINOLOGY ANNOTATIONS TO STANDARD VOCABULARIES

For reference purposes, we have mapped most of the annotations we've mentioned to well-known annotation properties and indicate their sources in Table 4.1.

Basic references that may be useful in developing your own annotations, some of which we reference in the table, include:

- ISO 1087, Terminology work — Vocabulary — Part 1: Theory and application;

- ISO/IEC 11179-3, Edition 3, Information technology — Metadata registries (MDR) — Part 3: Registry metamodel and basic attributes;

- OMG's Architecture Board Specification Metadata Recommendation; see http://www.omg.org/techprocess/ab/SpecificationMetadata.rdf;

- The Semantics for Business Vocabularies and Rules (SBVR) Specification from the OMG; see http://www.omg.org/spec/SBVR/1.4/;

- The Dublin Core Metadata Initiative (DCMI)'s Metadata Terms recommendation, see http://dublincore.org/documents/dcmi-terms/;

- The SKOS; see http://www.w3.org/2004/02/skos/; and

- The W3C's Provenance (prov-o) Ontology; see http://www.w3.org/TR/prov-o/.

Table 4.1: 1 Terminology to standard mapping

Element	Annotation	Definition	Source for Annotation
Concept	n/a	Unit of knowledge created by a unique combination of characteristics	ISO 1087-1:2000
Preferred label	prefLabel	A preferred lexical symbol or term for a resource, in a given language	W3C/SKOS

Alternate label	altLabel	An alternative lexical label for a resource	W3C/SKOS
Abbreviation	abbreviation	A short form for a designation that can be substituted for the primary representation	FIBO/FND, ISO 1087-1:2000
Community	community	Identifies the community that uses the concept and for which a particular label is preferred (or alternate)	
Synonym	synonym	An alternate designation or concept that is considered to have the same meaning as the primary representation in a given context	FIBO/FND, ISO 1087-1:2000
Synonym context	synonymContext	Identifies the community that uses the synonym	
Identifier	identifier	An unambiguous reference to the resource within a given context	dc/terms, ISO 11179-3:2013
Definition	definition	A representation of a concept by a descriptive statement which serves to differentiate it from related concepts	W3C/SKOS, ISO 1087-1:2000
Concept source	termOrigin	Document from which a given term was taken directly	FIBO/FND
Preferred label context	prefLabelContext	Identifies the community and/or policy that uses the preferred label	
Alternate label context	altLabelContext	Identifies the community and/or policy that uses the alternate label	
Definition source	definitionOrigin	Document from which a given definition was taken directly	FIBO/FND
Definition adaptation source	adaptedFrom	Document from which a given definition was adapted	FIBO/FND

Scope note	scopeNote	A note that helps to clarify the meaning and/or use of a concept [with emphasis on the scope of use]	W3C/SKOS
Explanatory note	explanatoryNote	A note that provides additional explanatory information about a given concept	FIBO/FND
Usage note	usageNote	A note that provides information about how the concept should be used in context	FIBO/FND
Example	example	An example of the use of the concept	W3C/SKOS
Dependencies	dependsOn	A bibliographic reference for and/or URL of any electronic files, documents, or other information source, including other concepts, on which the definition of this concept depends	OMG/SM
References	references	A related resource that is referenced, cited, or otherwise pointed to by the described resource	dc/terms
Concept status	conceptStatus	Rating established from a predetermined scale and used to evaluate a term	ISO 1087-1:2000
Concept status date	conceptStatusDate	Identifies the date a change to concept in the body of a vocabulary was made	
Steward	steward	Identifies the organization or person responsible for making a change to concept in the body of a vocabulary	
Change note	changeNote	A note describing a modification to a concept	W3C/SKOS

CHAPTER 5

Conceptual Modeling

In this chapter, we describe the primary steps involved in the conceptual modeling aspects of ontology development. While we give our examples using the family of languages known as description logic (DL), conceptual modeling may be done using other knowledge representation languages. We work primarily in description logic because: (1) the level of expressivity meets most of the requirements of the applications we build; (2) DL language design prioritizes support for efficient, sound, and complete reasoning; and (3) there are a number of open source and commercial tools available. This choice does not imply that DL alone is sufficiently expressive for every application, especially in cases where we need to execute rules or evaluate formulae. Most of the time, however, we can build ontologies in DL and augment them with rules or other logic for application purposes. Because many available tools support convenient checks for logical consistency of ontologies, we can eliminate many of the errors that would otherwise be inconvenient to find in large software and rule-based applications (Nitsche et al., 2014).

Our discussion of the formal aspects of description logic is limited to the basics needed in an introduction. For more in-depth coverage of a variety of topics related to knowledge representation in description logic, we recommend the *Description Logic Handbook* (Baader et al., 2003).

5.1 OVERVIEW

The modeling phase of an ontology engineering project requires, as inputs: one or more preliminary use case(s), a curated term list, and a business architecture and other business models if available, as discussed in Chapters 3 and 4. The use case(s) must include high-level requirements, preliminary competency questions, pointers to any known relevant vocabularies or ontologies, and references that may be used for domain analysis. The term list should be curated by SMEs, including initial definitions and metadata about the terms (e.g., the source(s) for definitions). If other sources for terminology are available, such as data dictionaries or interface definitions for data that is expected to be used in the application, terms used in those data dictionaries should be referenced by the curated term list. Given those inputs, an ontologist can develop a formal model that contains the concepts, their definitions, relationships between the concepts, and a set of formal axioms that distinguish the concepts from one another. This aspect of the work is *never* done in a vacuum, though. We work closely with SMEs to make sure that we can clearly and unambiguously tease out the distinctions between concepts, determine if we need to define new terms or reuse terms from existing ontologies, and represent the term characteristics and relationships at the right level of abstraction for the domain, community, and application.

Because an ontology is stated in a formal language with precise syntax and formal semantics, it is more accurate than a data dictionary or business glossary. Ontologies also provide more complete representations, at least at the conceptual and logical level. Relations between concepts, and constraints on and between concepts, are made explicit in an ontology, rather than implicit, which minimizes the chance of misunderstanding logical connections within the domain.

The kinds of ontologies described herein are structured as directed graphs to enable people to easily find relevant information, even when the terminology encoded in that graph is unfamiliar to them. They are also structured for automation, allowing them to be used to support searching and recommendation generation, business intelligence analyses, natural language processing, machine learning, and other data-centric applications. A well-formed ontology must enable every competency question to be asked and answered (including questions about processes and activities described by the use case), and cover enough of the ancillary vocabulary used in the use cases, term lists, capability maps and descriptions, and information maps and descriptions from the business architecture (if available) to meet requirements.

We cannot emphasize enough the need to work side by side with users, stakeholders, and SMEs to ensure that we capture the relationships and constraints correctly. People who come from a software or data engineering background sometimes assume that looking up definitions in dictionaries and modeling a domain based on what they can find online, or even in academic papers or corporate documents, is "good enough." That approach may work for certain research or software projects, but not for a serious ontology engineering effort. Typically, ontologies need to address both human communications and data governance and interoperability issues, which mandates understanding and modeling the domain from the stakeholders' perspectives. In other words, having a feedback loop built into the development process from the outset is essential to success. As a consequence, one of the first tasks is to arrange a review process. Ideally, we schedule regular meetings with SMEs to review the model as it evolves and prioritize next steps. We recommend weekly or twice weekly meetings in early stages of a project to jump-start the development. And, we find it best to start with a white board in a room with a few SMEs, rather than on a machine. If this much SME time is not available, then we set up a regular, if less frequent, review process that has as much domain input as is reasonable for the project.

The approach we take to modeling is based on many years' experience in domain analysis. There are several good references available that describe the process of conceptual modeling in description logic (Baader et al., 2003; Brachman et al., 1991b; Noy and McGuinness, 2001), which provide more detail in terms of the proof theory and underlying logic than we cover here. We use a combination of the UML (2017) and the Ontology Definition Metamodel (2014) for visual modeling, and the Web Ontology Language (OWL2) (Bao et al., 2012), which was under development when most of the DL references we cite were written, as the underlying description logic language. While the material we present in this chapter is based on recent work in OWL2, it can be generalized to any description logic. Our overall approach to ontology engineering is also applicable to ontology development in any concrete first-order logic language, such as the CLIF dialect of Common Logic (ISO/IEC 24707:2018, 2018).

Figure 5.1 provides an overview of the activities involved in ontology development, where the starting point "Develop Ontology" activity corresponds to the activity with the same name shown in Figure 5.1, where we present a more complete set of activities involved in knowledge engineering.

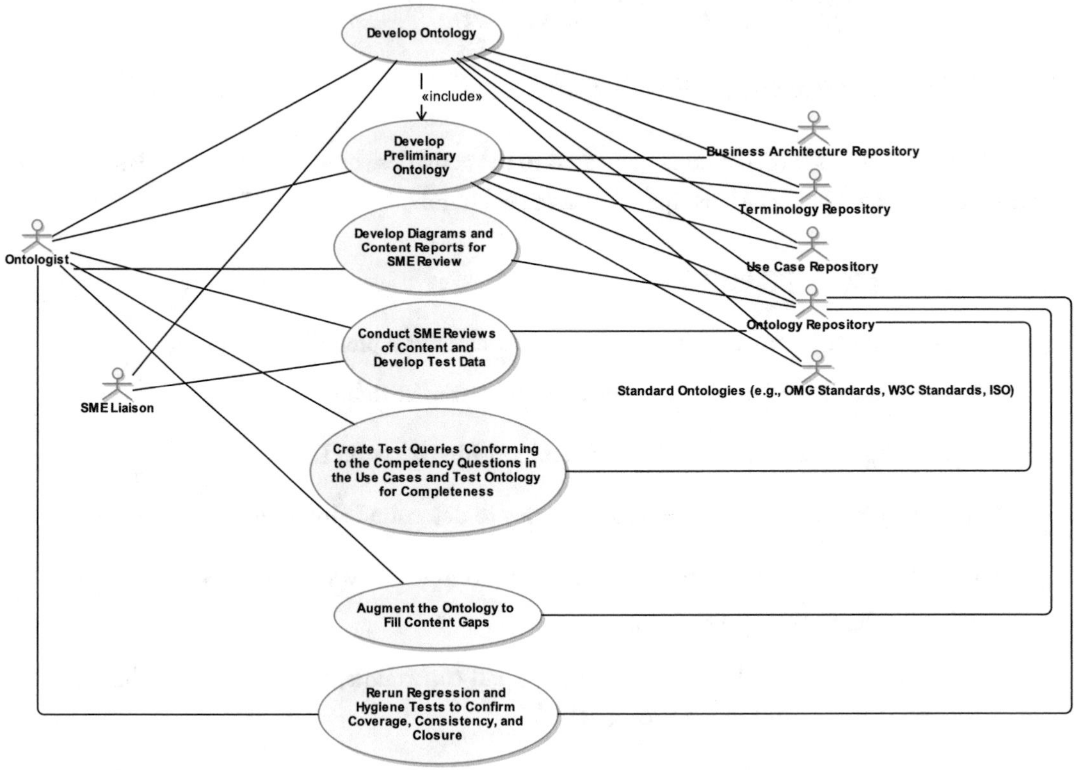

Figure 5.1: Ontology development activities.

Figure 5.2 expands on the preliminary development activity shown in the second bubble in Figure 5.1, providing insight into some of the more detailed work involved.

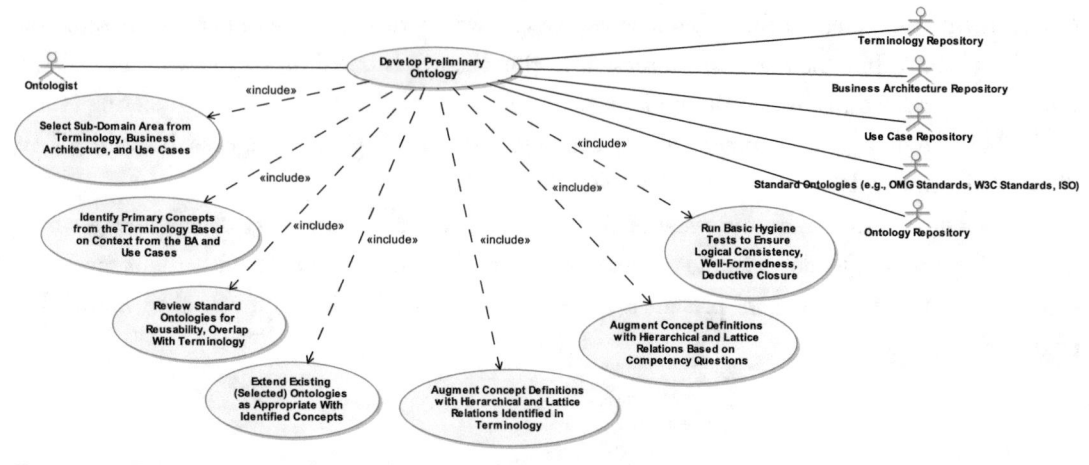

Figure 5.2: Preliminary ontology development steps.

The remainder of this chapter is dedicated to the activities shown in Figure 5.2, with emphasis on modeling concepts, relationships, and basic constraints.

5.2   GETTING STARTED

Figure 5.2 outlines the basic activities involved in developing a preliminary ontology:

- selecting a subset of a domain, use cases and term list to build out;

- identifying concepts and preliminary relationships between them;

- researching existing ontologies/vocabularies to determine reusability;

- identifying and extending concepts in existing ontologies/vocabularies (given that they are appropriate for our use case);

- connecting concepts through relationships and constraints, derived from (1) the term list and (2) competency questions; and

- conducting some basic tests to make sure what we have is consistent.

We typically select some small number of concepts as a starting point, flesh those out, perform preliminary testing that includes reviewing how these concepts support answering relevant competency questions, review the resulting model with SMEs, and then loop back and introduce additional concepts. As we iterate through this process, we revise the use case with respect to resources we identify (new actors), additional steps in our flows, and to add competency questions as we identify them. We also keep the terminology up to date as we identify new terms or revise existing metadata, such as resources, definitions, or other annotations. As we mentioned in Section

4.4, for one finance application we started with the concept of a *trade*, and through iterations that included developing competency questions involving trades, fleshing out the terminology corresponding to the competency questions, and conceptual modeling, we built out a graph that included more than 300 concepts. Some of these were atomic, such as trade date, settlement date, etc., while others, such as the concept of a deal, an offer, a financial instrument, a trading book, etc. were much more complex.

We work with stakeholders and SMEs to determine where to start, usually by identifying up to 20–30 primary terms from the term list that appear in high priority competency questions. We also organize the terms into rough topic areas that can be used to modularize the resulting ontology. There are some rules of thumb for modularization even very early in the development process, including limiting a given sub-graph, or subset of the ontology, to 100–150 concepts, if possible. For engineering applications, one might categorize designs by function, structure, and behavior, and classify documentation, processes, and other design artifacts independently, for example, separating disciplinary, process, manufacturing, evolution, and maintenance concerns. For medical applications, one can do something similar, using biological functions, medical disciplines (and capabilities within disciplines), and other perspectives to organize the content, representing data specific to those processes in parallel hierarchies. For finance applications such as those that leverage the FIBO, we've organized content by contract type (i.e., the nature of a financial instrument), by functional business area and capability (such as the capabilities carried out by a capital markets compliance organization), by reporting requirements, and so forth. Description logic languages are optimized to support multiple inheritance and multiple classification, which means that designing an ontology, or group of ontologies to "slice and dice" related data along many dimensions is not only supported but a best practice. Having a preliminary idea of how to structure and modularize the content at the start of an ontology project helps minimize reorganization and thrashing in downstream work.

5.3   IDENTIFYING REUSABLE ONTOLOGIES

There are many factors to consider with respect to selecting ontologies for reuse. Once we have identified the primary terms to model, we typically search well known repositories for relevant ontologies to consider. The University of Toronto maintains a list of some repositories, for example.[73] One of the most comprehensive repositories as of this writing is the BioPortal, maintained by the National Center for Biomedical Ontology and hosted by Stanford University.[74] BioPortal is focused primarily on medical and biological ontologies, but includes more general content, such as ontologies for scientific research, information artifacts, statistics, and quantities and units. The

---

[73] http://www.oor.net/.
[74] https://bioportal.bioontology.org/.

OMG has standardized several ontologies with more in the pipeline in finance, software engineering, aerospace, retail, and robotics domains, among others. The OMG standards are available via an online catalog.[75]

When we first starting building ontologies, there were very few that we could use as starting points, and no searchable repositories. That's no longer the case, however, and, as a result, we never start from scratch. We typically reuse ontologies for general metadata, as mentioned in Chapter 4, as well as for provenance and pedigree. For our students, depending on their use cases, we suggest topic-specific starting points in addition to the general-purpose metadata, and create seed ontologies that reference or import those starting points to jumpstart their development process. Some of our more common suggestions include:

- Dublin Core Metadata Initiative:[76] one of the most widely used vocabularies available today, developed by the library science community;

- SKOS: a W3C recommendation for very simple concept system representation, with useful metadata annotations even for use by those who are not developing SKOS vocabularies;[77]

- W3C PROV Ontology (PROV-O): a W3C recommendation that includes classes, properties, and restrictions for representing and exchanging provenance information;[78]

- Specification Metadata: an Object Management Group emerging standard for metadata about standards, including ontologies, with support for change management;[79]

- Nanopublications: a relatively recent approach to augmenting a knowledge graph with simple annotations for provenance;[80]

- Bibliographic Ontology: an ontology that supports source documentation[81] (there are several similar ontologies, but this is one of the more frequently cited ones); and

- Schema.org: Now an open community activity to create, maintain, and promote schemas for structured data.[82] With the emergence of data search services such as Google's

---

[75] https://www.omg.org/hot-topics/semantics-ontologies.htm.
[76] http://dublincore.org/.
[77] https://www.w3.org/2004/02/skos/.
[78] https://www.w3.org/TR/prov-o/.
[79] https://www.omg.org/techprocess/ab/SpecificationMetadata/.
[80] http://nanopub.org/wordpress/.
[81] http://bibliontology.com/.
[82] https://schema.org/.

dataset search,[83] using vocabularies that crawlers use is becoming more common. Google's dataset search will recognize schema.org annotations.

If we find one or more ontologies that are topically relevant, there are several additional things we look for before we recommend reuse. The most important of these is licensing. If the ontology in question, or the website that documents it, does not include copyright and licensing information, we will not even consider reusing it. Even in cases when we know the developers, if they are not willing to augment the ontology (or at least the website where they publish the ontology) with copyright and licensing information, such as a Creative Commons[84] or MIT license,[85] we cannot recommend the ontology for reuse and will not reuse it ourselves. We require our students to include copyright and licensing metadata in their ontologies and recommend this practice to our clients, even for internally published ontologies. It is up to the organization to determine which licenses are acceptable from a legal perspective.

Another serious consideration is maintenance. Anyone considering reusing an ontology should understand the maintenance policies of the developer. There are now hundreds of ontologies available on the web for potential reuse, including more than 700 from the BioPortal alone. The number of ontologies for which maintenance policies, such as update frequency, an indication of whether the developers make changes that modify the meanings of terms in ways that are inconsistent with definitions in previous releases, change history, and other change management details are available is far smaller. The number of BioPortal ontologies that have been maintained continuously and reused on other projects is also far smaller. No organization wants to invest in a significant ontology development effort that depends on some other ontology that might break at a moment's notice, or change in some subtle way that might cause downstream reasoning errors. The most frequently reused vocabularies and ontologies, such as the Dublin Core Metadata Terms, are popular partly because they are dependable. The Dublin Core Metadata Initiative publishes not only their policies for change management, but detailed change history for users to review. Many projects provide mechanisms for user feedback, including an automated facility for users to file issues and revision requests, such as a JIRA[86]-based approach. The most efficient teams also use a source code management system, such as GitHub[87] or Mercurial,[88] for tracking and managing the changes to their ontologies in a similar fashion that one would do for software configuration management. For the FIBO development work underway at the EDM[89] Council and OMG, custom scripts are used to serialize OWL2 ontologies consistently, and to perform automated unit tests prior to enabling

---

[83]   https://toolbox.google.com/datasetsearch/.
[84]   https://creativecommons.org/licenses/.
[85]   https://opensource.org/licenses/MIT.
[86]   https://www.atlassian.com/software/jira.
[87]   https://github.com/.
[88]   https://www.mercurial-scm.org/.
[89]   https://spec.edmcouncil.org/fibo/.

developers to formalize their changes through a "pull request," for example,[90] We recommend that our clients and students avoid reusing ontologies that do not use some sort of collaboration and change management tools for maintenance, and that they only reuse ontologies with an active user community committed to ongoing maintenance.

In addition to business considerations, such as licensing and maintenance, there are technical concerns to investigate prior to reusing an ontology. First and foremost, it must be syntactically correct, which can be verified using a validator, such as the RDF validation service from the W3C.[91] Second, it must be logically consistent. We check for logical consistency using DL reasoning engines such as HermiT,[92] Pellet,[93] and FaCT++.[94] For ontologies that include many qualified cardinality restrictions, data property restrictions and other constraints that may be difficult for complete reasoners to check in a short period of time, we may also use reasoners such as TrOWL,[95] which simulates time-consuming work performed by traditional tableaux reasoners such as HermiT and Pellet. Note that HermiT and Pellet are complete reasoners, that is, they iterate over the entire search space until no additional inferences can be made, whereas TrOWL is deliberately incomplete, and FaCT++ is also incomplete. We use at least one complete reasoner to determine whether or not an ontology is logically consistent.

Once we know that an ontology meets business requirements and is well-formed, we look for good topic coverage, meaning, a high percentage of reusable terms and definitions. One rule of thumb is that at least 50% of the terms in the ontology must be reusable/extensible, and limited conflict with terms and definitions identified by SMEs as important for the domain. Another consideration is modularization, meaning, whether or not the ontology is modularized in a way that a module with good domain coverage is easily separable from other components of the ontology that are not relevant to the use case. We also look for minimal completeness with respect to metadata— every concept and property must have a label and definition, and to the degree possible, a source for its definition. We have seen many cases in which people reuse an ontology simply because it includes one class or property they can extend or use, even if the ontology is quite large and makes other ontological commitments that they do not need or want. There is overhead (i.e., cost) involved in reusing any ontology. Even if an ontology is free to use and open source, potential users must understand the evolution of that ontology to ensure that it will continue to be applicable and won't

---

[90] See https://wiki.edmcouncil.org/display/FIBO/FIBO+Build+and+Publish+Process for details about the approach.

[91] https://www.w3.org/RDF/Validator/.

[92] HermiT OWL Reasoner from the Information Systems Group in the Department of Computer Science at the University of Oxford; see http://www.hermit-reasoner.com/.

[93] https://github.com/stardog-union/pellet.

[94] FaCT++ Reasoner from the Information Management Group in the School of Computer Science at the University of Manchester.

[95] See https://www.semanticscholar.org/paper/TrOWL%3A-Tractable-OWL-2-Reasoning-Infrastructure-Thomas-Pan/2f748d5684a5d82fdc0fb3b800c77033686971e4 for a description of the TrOWL reasoner.

change in a destructive manner with respect to their application, for example. By destructive, we mean any change that either (1) deletes any concept or property rather than deprecating it, or (2) revises the logic in ways that make the ontology logically inconsistent with prior versions.

In practice, we tend to reuse foundational ontologies, such as those for metadata, basic knowledge such as time and units of measure, or core domain-level content ontologies, for example, top-level ontologies from FIBO, more often than narrow but deep, detailed ontologies developed from a specific point of view. There tend to be more issues, from a quality and sometimes licensing perspective, the more specific the ontologies get. In some cases, where deeper and narrower ontologies are heavily constrained (i.e., very specific), the commitments made via logical axioms may not be true for all applications. This does not mean, for example, that we would not recommend reuse of well-tested and highly detailed biomedical ontologies, but we suggest thorough analysis prior to reuse of any ontology to ensure that it meets your requirements. This is another place where competency questions play a role—do you get the answers you expect when you pose a representative competency question against a candidate ontology for reuse? Is that still the case if you integrate it with other ontologies you have developed or have elected to reuse?

For our landscape example, one potential reference ontology is the *National Agricultural Library Thesaurus* (NALT),[96] The NALT thesaurus is published as linked data, and includes close to 65,000 individuals. It is limited with respect to ontological commitments, however and is modeled as a SKOS vocabulary that may be difficult to navigate depending on tool selection. There are no classes or hierarchical relationships in the model, although there are many broader/narrower relations between individuals. The plant science part of the thesaurus provides reference data, however, and is a good source of definitions. There are several ontologies in the BioPortal related to plants, each of which has distinct coverage and a different community of interest, ranging from agricultural research to biodiversity and biomedical research. Several of these are quite large, such as the *NCI Thesaurus*, which contains definitions that are relevant to our use case. From a pragmatic perspective, we would likely choose not to import either ontology directly, but would reference them both as sources for definitions. One practice that we use when ontologies are large but we want to leverage a small percentage of the ontology content is to use a "MIREOT" approach.[97] This stands for the minimum information to reference an external term. Using this approach, we just reference the external term, maintaining metadata about the source.

5.4   PRELIMINARY DOMAIN MODELING

The next step in the process is to begin modeling. Typically, when we build seed ontologies as the starting point for a project, we create an "empty" model with the name of the target ontology, and

---

[96] https://agclass.nal.usda.gov/dne/search_sc.shtml.
[97] http://preceedings.nature.com/documents/3574/version/1.

import (load) the set of reusable ontologies we've identified as a starting point. Given that seed model, we locate or create the 20–30 primary concepts identified with SMEs from the term sheet, including the metadata about those concepts (e.g., label, definition, source, other notes). Concepts identified in the term list become classes in our model. For a trading system, these might include: trade, trader, trading platform, financial instrument, contract, party, price, market valuation, etc. In a nursery example, the classes might be: container, hardscape material, landscape, light requirement, plant, planting material, and so forth. At this stage, we also model any structural relationships between these primary concepts that we are aware of, including hierarchical relationships and lateral relationships. By hierarchical relationship, we mean specialization/generalization relations, commonly called parent/child relationships. In an ontology, these must always be true "is-a" relations, or the logic will not work as anticipated. Lateral relationships might include whole-part or other mereological (i.e., part-like) relations, membership and other set theoretic relations (e.g., x is a member of y), or other functional, structural, or behavioral relations that we can identify from the definitions of the concepts. We also include any known key attributes, such as dates and times, identifiers, and the like.

We have used various tools for modeling purposes over the course of our respective careers, but recommend visual tools at this point in the process. As mentioned at the start of Chapter 2, our goals include:

- "drawing a picture" that includes the concepts and relationships between them, and

- producing sharable artifacts, that vary depending on the tool—often including web sharable drawings.

As an example, let's start with the user stories from the landscape use case in Chapter 3:

- *I would like to purchase flowering plants that should do well in sunny areas of my garden.*

- *I am designing a south-facing bed that is partially shaded, and would like drought-tolerant plant suggestions.*

- *I would like plants that are not likely to attract and be eaten by deer.*

- *I need to purchase a new tree that is oak-root fungus resistant to replace one that recently died of that disease. It must be selected from the city of Los Altos preferred tree list, and of those, I would like one that flowers and whose leaves turn color in the fall.*

Primary terms from these stories include plant, flowering plant, garden, tree, flowering tree, deciduous tree, sunny, partially shaded, south-facing, drought-tolerant, deer resistant, and oak-root fungus disease resistant, with the City of Los Altos preferred tree list[98] and the University of California at Davis' Home and Garden Resources[99] as a source.

To create a seed ontology, we will incorporate the main metadata ontologies that we typically use, including Dublin Core, SKOS, and the Specification Metadata from the OMG as starting points, and will reference the two vocabularies mentioned above for definitions and for individuals, as appropriate. We also need a small botanical taxonomy as the basis for describing cultivated plants. Several ontologies in the BioPortal incorporate taxonomic terminology, including the Biological Collections Ontology (BCO) and Plant Experimental Assay Ontology (PEAO), which references the Darwin Core Terms.[100] Darwin Core meets many of our requirements listed above, but is strictly an RDF vocabulary, and the other ontologies are far too large, so we will link our definitions to them or to the NALT and NCI thesauri in our seed ontology, as appropriate.

5.5 NAMING CONVENTIONS FOR WEB-BASED ONTOLOGIES

Naming conventions and versioning policies are critical for every software engineering, data management and governance, and ontology development project, including policies for management of the resulting artifacts. Because our ontologies are designed to be used on the Web, care should be taken to ensure that IRI naming and versioning approaches (1) conform to the standards and best practices published by the W3C, (2) conform to organizational policies, and (3) are sufficiently extensible to accommodate changes as the number and nature of the ontologies in our ecosystem grows. Once something has been published with a particular IRI, the resource at that IRI should never change. In other words, if you post something at an IRL, the idea is that it should be permanent, so namespace design and governance is essential and should be decided well in advance of publication. New versions should be published at new versioned IRIs rather than overwriting prior versions, primarily so that we do not disenfranchise our user community. People need to know that they can trust what we've published and that it won't change, but also that they can find the latest version if they are ready to update the applications that depend on it. This means that publishing your policies for maintaining an ontology are also important, including indicating something about the frequency with which you think you will be updating it.

A common pattern for version specific ontology IRIs is:

```
https://<authority>/<business unit>/<domain>/<subdomain>/<date, in YYYYMMDD
form, or some other versioning strategy>/<module>/<ontology name>/
```

where

[98] https://www.losaltosca.gov/sites/default/files/fileattachments/Building%20and%20Planning/page/432/streettreelist.pdf.

[99] https://www.plantsciences.ucdavis.edu/plantsciences/outreach/home_garden.htm.

[100] http://rs.tdwg.org/dwc/terms/.

- **authority:** the top level corporate or other organizational domain, e.g., spec.edmcouncil.org, or ontologies.tw.rpi.edu;

- **business unit:** optional, but for many corporate projects is required;

- **domain:** this might be a topic area or possibly the top-level in the IRI hierarchy for a project;

- **subdomain:** optional, but often a further refinement from the topic or project level;

- **version:** this depends typically on the approach taken by an organization to versioning, noting that often an organization will use content negotiation to link the most current version of an ontology to a non-versioned IRI for convenience for users;

- **module:** we recommend modularizing ontologies to the degree possible so that a given application only needs to load what is relevant; and

- **ontology name:** the actual name of the ontology; the corresponding file name will include the ontology name with an appropriate extension depending on the language and serialization; common extensions for RDF or OWL ontologies include `.rdf` for RDF/XML serializations or `.ttl` for a Turtle serialization, or others depending on the ontology language and concrete syntax selected.

Levels of hierarchy may be inserted in the IRI structure for large organizations or to reflect modularization strategies, and of course, not all of them are required for small projects in smaller organizations.

Namespace prefixes (abbreviations) for individual modules modules and ontologies, especially where there are multiple modules, are also important. Certain prefixes, such as for the Dublin Core, SKOS, the W3C Prov-O ontology, and many others are commonly used and people tend to remember them over the IRIs. We often recommend an approach for prefixes that mimics the IRI structure. For example, there are many ontologies that make up the FIBO, with a number of top-level domains and lower level modules. For the FIBO ontologies published by the OMG, a typical versioned IRI is structured as follows:

```
https://www.omg.org/spec/EDMC-FIBO/<domain>/<YYYYMMDD>/<module>/<ontology
name>/
```

For a recent release the version IRI for one of the ontologies is:

```
https://www.omg.org/spec/EDMC-FIBO/FND/20180801/Arrangements/Classification-
Schemes/
```

The corresponding ontology prefix is

```
<fibo>-<fnd>-<arr>-<cls>
```

In other words, the prefix is structured:

```
<spec>-<domain>-<module>-<ontology abbreviation>
```

We bring this up now because, often when you use tools to create an ontology, the first thing you need to do is give it a name, IRI, and a corresponding namespace prefix.

Best practices in namespace development can be summarized as follows.

- **Availability:** people should be able to retrieve a description about the resource identified by the IRI from the network (either internally or externally to their organization and depending on whether the ontology is published on an intranet or publicly).

- **Understandability:** there should be no confusion between identifiers for networked documents and identifiers for other resources.

    ◦ IRIs should be unambiguous;

    ◦ IRIs are meant to identify only one of them, so one IRI can't stand for both a networked document and a real-world object; and

    ◦ separation of concerns in modeling "subjects" or "topics" and the objects in the real world they characterize is critical, and has serious implications for designing and reasoning about resources.

- **Simplicity:** short, mnemonic IRIs will not break as easily when shared for collaborative purposes, and are typically easier to remember.

- **Persistence:** once an IRI has been established to identify a particular resource, it should be stable and persist as long as possible.

    ◦ Exclude references to implementation strategies, as technologies change over time (e.g., do not use ".php" or ".asp" as part of the IRI scheme), and understand that organization lifetime may be significantly shorter than that of the resource.

- **Manageability:** given that IRIs are intended to persist, administration issues should be limited to the degree possible:

    ◦ some strategies include inserting the current year or date in the path so that IRI schemes can evolve over time without breaking older IRIs; and

    ◦ create an internal organization responsible for issuing and managing IRIs, and corresponding namespace prefixes if such a governance role does not already exist.

Once we have a namespace and prefix established for our ontology, we need to think about naming the first class elements in our ontology (concepts, relationships, individuals). Many organizations have naming conventions in place for data modeling and software engineering activities, which tend to vary by community of practice. Data modelers often use underscores at word boundaries and spaces in names, which not all tools handle well, and which result in intervening escape characters, such as "%" in IRIs. Some modelers name properties to include the domain and range with respect to a predicate (source and target classes), although not necessarily consistently. Because properties are first class citizens in an ontology, names that include the source and target *limit the reusability* of a property, thus limiting the ability to use common queries across multiple resources that contain similar relationships. Goals with respect to ontology element naming include human readability, elimination of ambiguity, and use with the broadest range possible of tools. Some of the potential tools that should be considered include not only modeling tools, but knowledge graphs and underlying stores, data profiling tools, information extraction and natural language processing tools, and machine learning applications. To meet these goals, particularly for interoperability, semantic web practitioners typically use camel case (i.e., CamelCase) for naming without any intervening spaces or special characters. IRIs tend to be primarily lowercase, whereas ontology and element names tend to be camel case: upper-camel case for class and datatype names, lower-camel case for properties. Use of verbs to the degree possible for property naming, without incorporating a domain or range (source or target), maximizes reuse and is a best practice for ontologies designed to support natural language processing (NLP) applications, among others. For very large ontologies, such as some in the biomedical informatics community or for automatically generated ontologies, many people use unique identifiers to name concepts and properties, with human readable names in labels only. This approach works well for large code lists and identifier schemes, but may be less useful in the case of a business vocabulary intended to support data governance, again depending on the target application environment. The key is to establish and consistently use guidelines tailored to your organization.

5.6   METADATA FOR ONTOLOGIES AND MODEL ELEMENTS

Metadata should be standardized at the ontology level and the element level. Ontology level metadata can reuse properties from the Dublin Core Metadata Terms, from the SKOS, ISO 11179 Metadata Registry standard with ISO 1087 Terminology support, W3C PROV-O vocabulary for provenance and others, such as the examples provided in Table 4.1. Most annotations can be optional at the element level, especially when a project is just getting underway, but a minimal set, including names, labels, and formal definitions with their source and any required approval information is important for reusability and collaboration.

Consistent use of the same annotations and same annotation format (properties, tags) improves readability, facilitates automated documentation generation, and enables better search over ontology repositories. Some organizations have found improved search engine results through the use of tags (microformats, RDFa, JSON-LD) that are generated from their ontologies and embedded in the relevant web pages.[101] Tagging internal policies, procedures, and other important documents using the terminology specified as an ontology representing the preferred business vocabulary for data governance may significantly reduce the time people spend searching for those documents, in addition to supporting operational goals for applications that leverage the ontology.

5.7   GENERAL NATURE OF DESCRIPTIONS

Our goal in designing any ontology, regardless of the language or syntax used, is to describe the important concepts in a domain, the relationships between them, and to articulate axioms that distinguish those concepts from one another. The concepts identified in the use cases and term sheets provide the starting point for the analysis required to do this. For example, in an ontology that defines terminology for wine and food, we might start by defining the concept of win,e as shown in Figure 5.3.

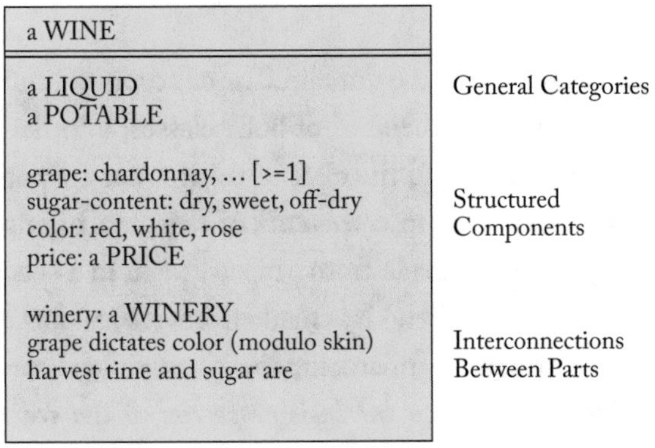

a WINE	
a LIQUID a POTABLE	General Categories
grape: chardonnay, … [>=1] sugar-content: dry, sweet, off-dry color: red, white, rose price: a PRICE	Structured Components
winery: a WINERY grape dictates color (modulo skin) harvest time and sugar are	Interconnections Between Parts

Figure 5.3: The general nature of wine.

In the figure, wine is defined as being a member of two general categories of things—as a liquid, and as something potable. We also specify some properties of wine that distinguish it from other liquids and potable things, such as that it is made from grapes (or another fruit or edible substance, potentially), that it has sugar content that varies over a controlled vocabulary, that it has color that again varies per a controlled vocabulary, and that it may have a price associated with it.

---

[101]   OMG has recently revised its Specification Catalog to tag the specification pages automatically using JSON-LD and based on the specification metadata ontology published by the Architecture Board. Resulting searches for specifications including UML, BPMN, SysML, and others found the tagged pages quickly and those pages tend to jump to the top of a search results list over other books and resources as a result, based on comparisons of statistics for those pages before and after the tags were added. See https://www.omg.org/tech-process/ab/specificationmetadata/1.0/SpecificationMetadata.rdf.

There are additional things that we can say about it, such as to identify the winery that made it, and to relate other properties, such as late harvest, with sweeter wines.

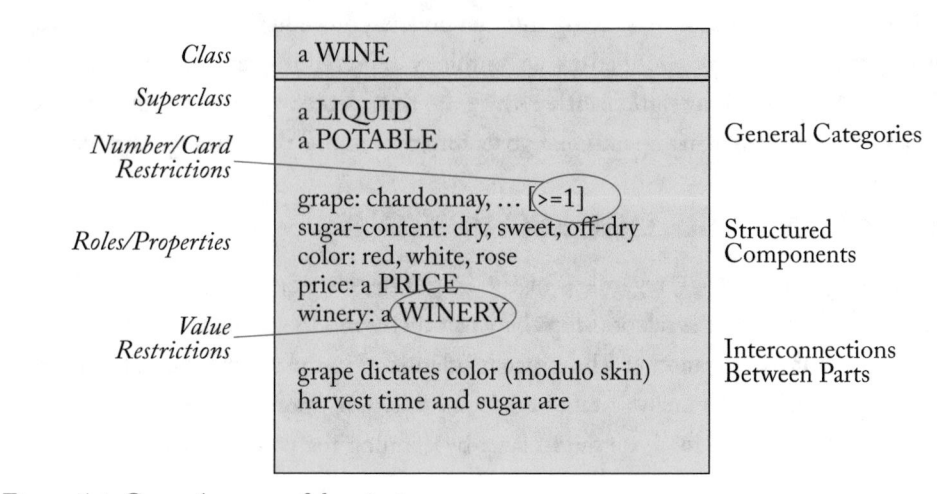

Figure 5.4: General nature of descriptions.

A class corresponds to a concept in the domain. The concepts of liquid and potable, as shown, are classes in the domain, and wine is a member of both classes; in other words, the wine class is in the *intersection* of the class of liquids and the class of things that are potable. The class of liquids and the class of things that are potable are *superclasses* of the class of wines. Other classes in the domain might include vintage—a wine made from grapes grown in a specified year. The structural characteristics of wine, such as which grape(s) it is made from, it's residual alcohol content, its color, whether it is considered sweet or dry, the microclimate where the grapes were grown, the winery that produced it, etc. are characteristics, or *intrinsic properties* of the wine. They are fundamental to the nature of the wine. Other characteristics, such as the suggested distributor's retail price, the color of the bottle, the image on the label, are extrinsic properties of the wine, i.e., they are imposed externally. A class is a collection of elements with similar characteristics, such as white wine is a class of wines made without processing the skins; white table wine made from grapes that are not appellations or regional (not "quality wine" in the EU). A class expression defines necessary and sufficient conditions for membership, such as a specific varietal, that the grapes were grown in a specific field in a given vineyard which has a specific microclimate, that the grapes were picked on a certain date, that the wine was bottled on a certain date, and so forth. These expressions are built up from a combination of properties and classes using additional logic.

Instances (or individuals) of classes in this particular domain include:

- Ridge 2012 Kite Hill Zinfandel, York Creek Vineyard, Appellation Napa,[102] 86% Zinfandel, 14% Petite Syrah;

- Howell Mountain American Viticultural Area (AVA), Napa Valley, California; and

- Tracy Hills AVA, within BOTH San Joaquin County and Stanislaus County, California.

Some properties relate the wine class to other classes, such as to the winery, the vineyard, the individual wine maker, the viticultural area, etc. We consider these kinds of properties to be first class relationships, or "roles" in description logic ("object properties" in OWL). Others, such as the residual alcohol content, might be modeled as attributes in a data model, or as "data properties" from an OWL perspective, whose values are expressed as data types (strings, dates, numbers, etc.). In both cases we can say more about the relationships, such as that in order to be a member of the wine class, it must be made from *at least one* kind of grape. "At least one" is called a *number* or *cardinality* constraint. Saying that wine must be made by a something that is categorized as a winery is called a *value restriction* in description logic. In general, these kinds of modeling choices are not always straightforward: there are often multiple ways of representing the same thing. The competency questions and scope specified in the use case can help determine the best way to model something from the perspective of the application(s) that the ontology is intended to support.

Classes are organized into sub-class/super-class hierarchies (also called specialization/generalization or hyponymic relations), in an ontology. These relationships are the basis of a formal *is-a* hierarchy:

- classes are is-a related if an instance of the subclass is an instance of the superclass;

- is-a hierarchies are essential for classification; and

- violation of true is-a hierarchical relationships can have unintended consequences and will cause errors in reasoning.

Class expressions should be viewed as sets, and subclasses as a subset of the superclass. Examples of class expressions from our garden example include:

- FloweringPlant is a subclass of Plant;

  Every FloweringPlant is a Plant; every individual flowering plant (e.g., Azalea indica "Alaska" (Rutherfordiana hybrid) is an instance of FloweringPlant.

- Azalea is a subclass of FloweringPlant and Rhododendron; and

- Monrovia is a company that breeds, grows, and sells flowering plants.

---

[102] https://www.ridgewine.com/wines/2012-kite-hill-zinfandel/.

Note that we have worked with people who, as a matter of convenience, have modeled collections of sometimes orthogonal or completely independent attributes as classes in object-oriented programming languages, and then "inherited" from these classes as they saw fit in their programs. Such collections typically do not identify true sets, however, whose subclasses each contain subsets of the members of the superclass. The likelihood of causing reasoning errors when taking these kinds of shortcuts is high.

Practitioners tend to approach modeling in various ways, depending on the competency questions, resources (including subject matter expert availability), and target prototype or application requirements.

- Top-down: define the most general concepts first and then specialize them.

- Bottom-up: define the most specific concepts and then organize them into more general classes.

- A combination of the two approaches: breadth at the top level and depth along a few branches to test design.

It is also important to understand that inheritance is transitive: if A is a subclass of B and B is a subclass of C, then A is a subclass of C. For example, consider B1 to be "white wine," and B2 to be "dessert wine," both of which are subclasses of C, "wine." Then consider A1 to be "sauvignon blanc wine," which is a subclass of B, "white wine," and A2 to be "late harvest wine," which is a subclass of B2, "dessert wine." Then A1 and A2 are also subclasses of C, "wine." Further, a class such as "late harvest sauvignon blanc wine" is a subclass of both A1 and A2, and therefore is also a subclass of both B1 and B2, and ultimately a subclass of C.

In this example, we could have started modeling from multiple places in the hierarchy—at the bottom, with "late harvest sauvignon blanc wine," in the middle, with either its direct parents or grandparents, or at the top, from "wine," depending on the artifacts and competency questions we were attempting to support. We typically use a combination approach, starting from the top and building out a relatively broad 2–3 level hierarchy, laying out a high-level architecture for key ontologies and the classes that they include. From there, we focus on identifying the relationships among those classes, teasing them out in a more bottom-up fashion. Additional classes may be differentiated based on whether they are core to the domain or a subset of the domain, whether they are specific to certain interfaces or processes from a systems perspective, whether we need them as foundational concepts at higher levels in the hierarchy to fill gaps we identify along the way, and so on.

There are a few additional tools we want to mention that some practitioners use for capturing concepts and getting a jump on the conceptual modeling phase of a project that are particularly useful on larger projects. These include IDEF (Integrated Definitions Methods) 5—Ontology

Definition Capture Method[103] and use of IDEF 5 style tools for gathering content. Another approach uses an OMG standard, SBVR (Semantics for Business Vocabularies and Rules[104]) for understanding and documenting concepts and their relationships. IDEF5 in particular provides a methodology for gathering information from SMEs, that includes:

- understanding and documenting source materials;

- an interview template; and

- traceability back to your use cases.

For every class in the ontology, it provides a mechanism for:

- describing its domain and scope;

- identifying example questions and anticipated/sample answers for the application(s) it will support;

- identifying key stakeholders, ownership, maintenance, resources for corresponding individuals and related data sources;

- describing an anticipated reuse/evolution path; and

- identifying critical standards, resources that it must interoperate with, and dependencies.

Most of these things are already covered in our use case template, but IDEF5 takes this to another next level of specificity, and provides some additional documentation templates and methods that may be useful for large projects.

5.8   RELATIONSHIPS AND PROPERTIES

Relationships and properties are used to further refine class definitions in any ontology. They typically describe the characteristics and features of the members of a class, as well as link that class to others in a knowledge graph. For example, from our landscape use case:

> *Every flowering plant has a bloom color, color pattern, flower form, and petal form.*

Properties can be categorized further, being *intrinsic* or extrinsic, *meronymic* (partonomic), representing *geospatial* or *temporal* aspects of something, cause and effect relations, identity and ownership relations, and so forth (Green, Bean, and Myaeng, 2002). An intrinsic property is a property reflecting the inherent nature of something, for example, for a flowering plant, intrinsic properties include the form its leaves or petals take, optimal sunlight and soil conditions for proper growth, and its expected height at maturity. An extrinsic property is one imposed externally, such

[103]   http://www.idef.com/idef5-ontology-description-capture-method/.
[104]   https://www.omg.org/spec/SBVR/.

as the grower and retail price. Meronymic relations, often called part-whole relations in artificial intelligence, are fundamental relationships that can be as important as inheritance relations in an ontology. Spatio-temporal relationships, including geospatial and temporal relations, may also be important depending on the domain and application. In our examples, some of the more important geospatial relationships include relating a plant to the microclimates it is well-suited to, which USDA plant hardiness zones[105] it is most likely to thrive in, and for large, historic gardens or arboretums, where one can find a specimen.

In description logic, properties are further differentiated as object properties (relationships between classes and individuals that are members of those classes) and data properties (traditional attributes from an entity relationship diagram perspective). Data properties are used for simple attributes that typically have primitive data values (e.g., strings and numbers). Complex properties (relations) refer to other entities (e.g., a retailer/grower, such as Monrovia,[106] or association, such as the American Orchid Society[107]). Description logic reasoning requires strict separation of datatypes (primitives, using the XML schema datatypes in the case of OWL) and object (individuals of complex classes), and related to that, data and object properties, in order to maintain decidability (Baader et al., 2003).

In description logic, relations (properties) are strictly binary. In other knowledge representation languages, such as Common Logic (2018), *n-ary* relations are inherently supported. However, there are a number of common patterns for mapping *n-ary* relations to binary relations[108] that are useful to get around this. Most *n-ary* relations are ternary or, less frequently, quaternary relations, for example, used to represent indirect objects of certain verb phrases. For binary relations, the *domain* and *range* are the source and target arguments of the relation, respectively. These concepts are similar to their definitions in mathematics: the domain is the set of individuals or values for which a relation is defined (i.e., if a relation, or role in description logic, is defined as a set of tuples of the form (x, y), then the domain corresponds to all individuals or values of x). Similarly, the range is the set of individuals or values that the relation can take, i.e., the individuals or values of y. Another way to look at domain is as the class, or classes that may have a property, such as, *wine is the domain of hasWineColor*, and the range as the class or classes that every value that fills the property must instantiate , such as the enumerated class, {*red, white, rose*}. Some KR languages that inherently support *n-ary* relations, such as Common Logic (CL), do not make this distinction. The formalisms these kinds of languages use tend to be more flexible, where functions have ranges (or return types) but not all relations are functions. Specifying *n-ary* relations in these languages may require additional detail to specify argument order, which can be critical for ontology alignment.

---

[105] https://planthardiness.ars.usda.gov/.
[106] https://www.monrovia.com/.
[107] http://www.aos.org/.
[108] https://www.w3.org/TR/swbp-n-aryRelations/.

Description logic provides a number of ways that we can constrain property values further. Number restrictions can be used to either require or limit the number of possible values that can occur in the range of a property, for example:

- a natural language as defined in ISO 639-1 has at least one English name and at least one French name[109] and

- a natural language may optionally have indigenous (local) names.

Cardinality, which is defined in description logic as in classical set theory, measures the number of elements in a set (class).

- Cardinality: cardinality N means the class defined by the property restriction must have exactly N values (individual or literal values). Thus, it must have at least N values and at most N values—thus exactly N values.

- Minimum cardinality: 1 means that there must be at least one value (required), 0 means that the value is optional.

- Maximum cardinality: 1 means that there can be at most one value (single-valued), N means that there can be up to N values ($N > 1$, multi-valued).

Figure 5.5 shows a class of individuals that have exactly one color (single colored), and a class of individuals that have at least two colors (multicolored). Note that the relationship between *SingleColoredThing* and the restriction class is modeled as an ***equivalence relationship***, meaning that membership in the restriction class is a ***necessary and sufficient condition*** for being a *SingleColoredThing*.

A qualified cardinality restriction allows us to not only restrict the number of elements in the range of a given property, but to refine the range class or datatype itself, typically to one that is a subclass of the original range of the property, or a datatype that refines the datatype in the range of the property.

---

[109]   See https://www.omg.org/spec/LCC/ for ontologies representing ISO 639-1 and 639-2 Language Codes, and ISO 3166-1 and 3166-2 Country and Country Subdivision Codes.

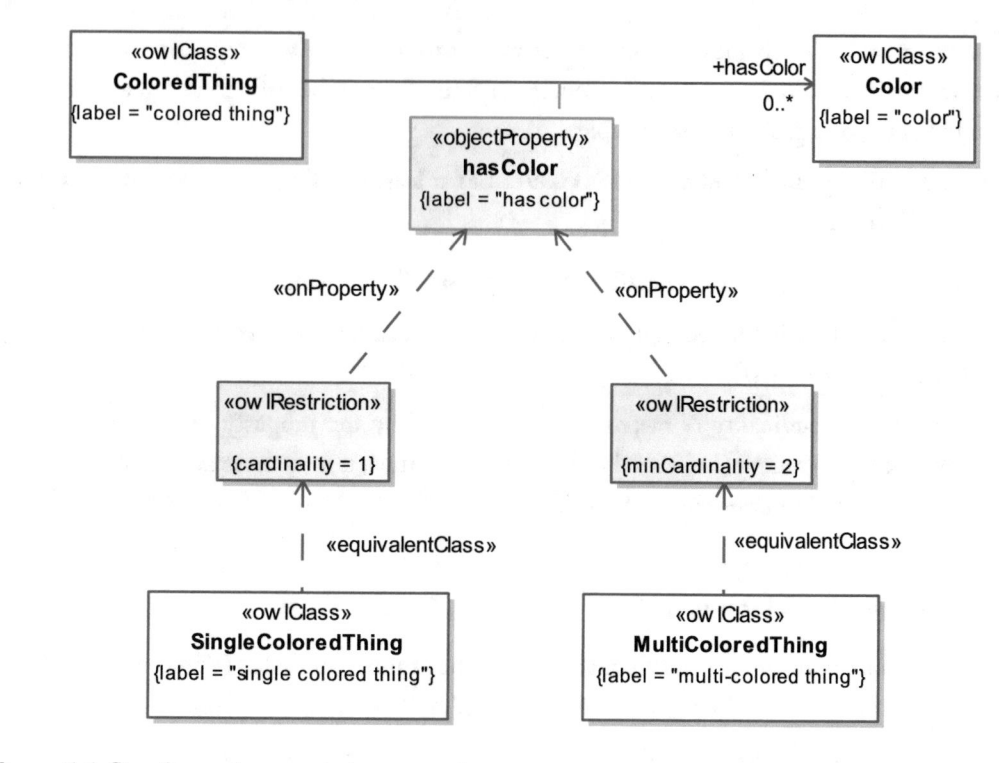

Figure 5.5: Simple number restriction example.

Patterns like the one shown in Figure 5.6, which reuse a smaller number of significant properties, building up complex class expressions that refine definitions of subclasses, are common in ontologies designed to support data classification and categorization, query answering, recommendation-making, systems configuration, natural language processing, and other applications.

Description logic provides other mechanisms for specifying concept definitions, such as through class expressions that restrict the possible values for properties by type using quantification. *allValuesFrom* is used for "*for all x*" ($\forall(x)$), or universal quantification, and restricts the possible values of a property to members of a particular class or data range. *someValuesFrom* is used for "*there exists an x*" ($\exists(x)$), or existential quantification, and restricts the possible values of a property to *at least one* member of a particular class or data range. *hasValue* restricts the possible values of a property to either a single data value (e.g., the *hasColor* property for a *RedWine* must be satisfied by the value "*red*"), or an individual member of a class (e.g., an individual of type *Winery* must satisfy the value restriction on the *hasMaker* property on the class, *Wine*). Enumerations, or sets of allowable individuals or data elements, can also be used to constrain the possible values of properties. We often use combinations of class expressions to build up richer models of some concept, including restrictions as well as Boolean expressions (*union*, approximating *logical or*, *intersection*, approximating *logical and*, and less frequently, *complement*, approximating *logical not*).

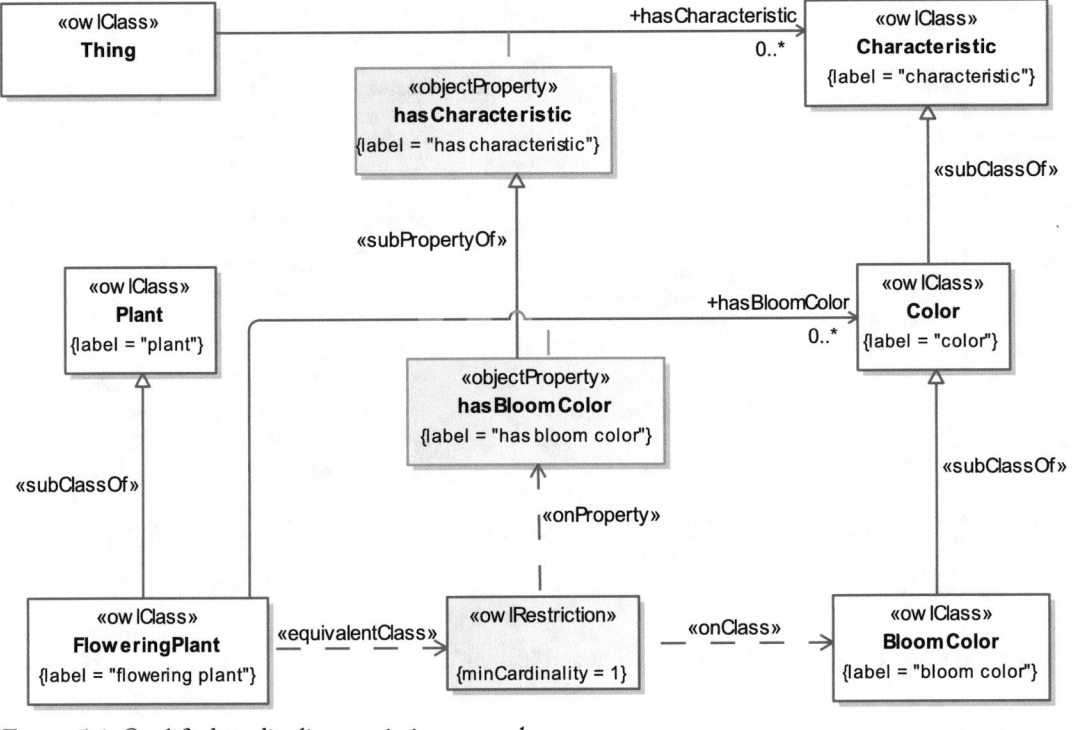

Figure 5.6: Qualified cardinality restriction example.

Figure 5.7 can be interpreted as follows.

> *Azaleas are flowering plants whose bloom color must be either an RHSColor (Royal Horticultural Society) or ASAColor (Azalea Society of America).*

We often use combinations of class expressions to build up richer models of some concept, including restrictions, as well as Boolean expressions (*union*, approximating *logical or*, *intersection*, approximating logical and, and less frequently, *complement*, approximating *logical not*).

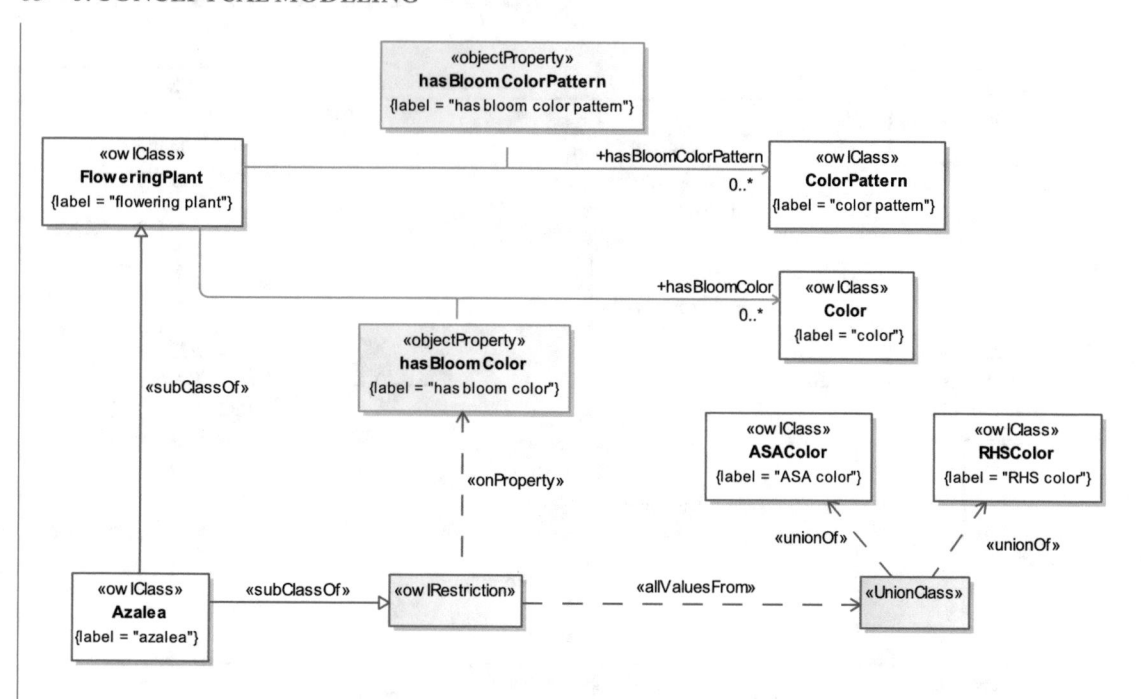

Figure 5.7: Class expression using universal quantification.

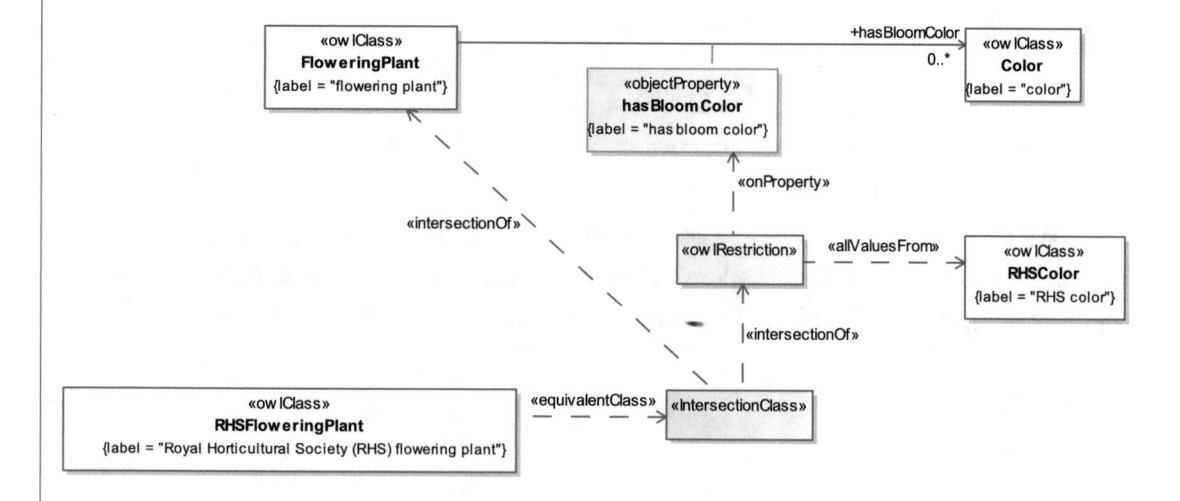

Figure 5.8: Class expression combining restrictions and Boolean connectives.

In Figure 5.8, *RHSFloweringPlant* is defined as being equivalent to a *FloweringPlant and* having a bloom color whose value is an individual of type *RHSColor*. Even if we don't explicitly state that *RHSFloweringPlant* is a *FloweringPlant* through an explicit subclass relation, a description logic reasoner would infer that it is.

Figure 5.9 provides an example of a class expression that uses both existential quantification and exact values for properties.

Figure 5.9 can be interpreted asfollows.

> *A SingleColoredAzalea is an Azalea that has precisely one bloom color and a solid color pattern.*

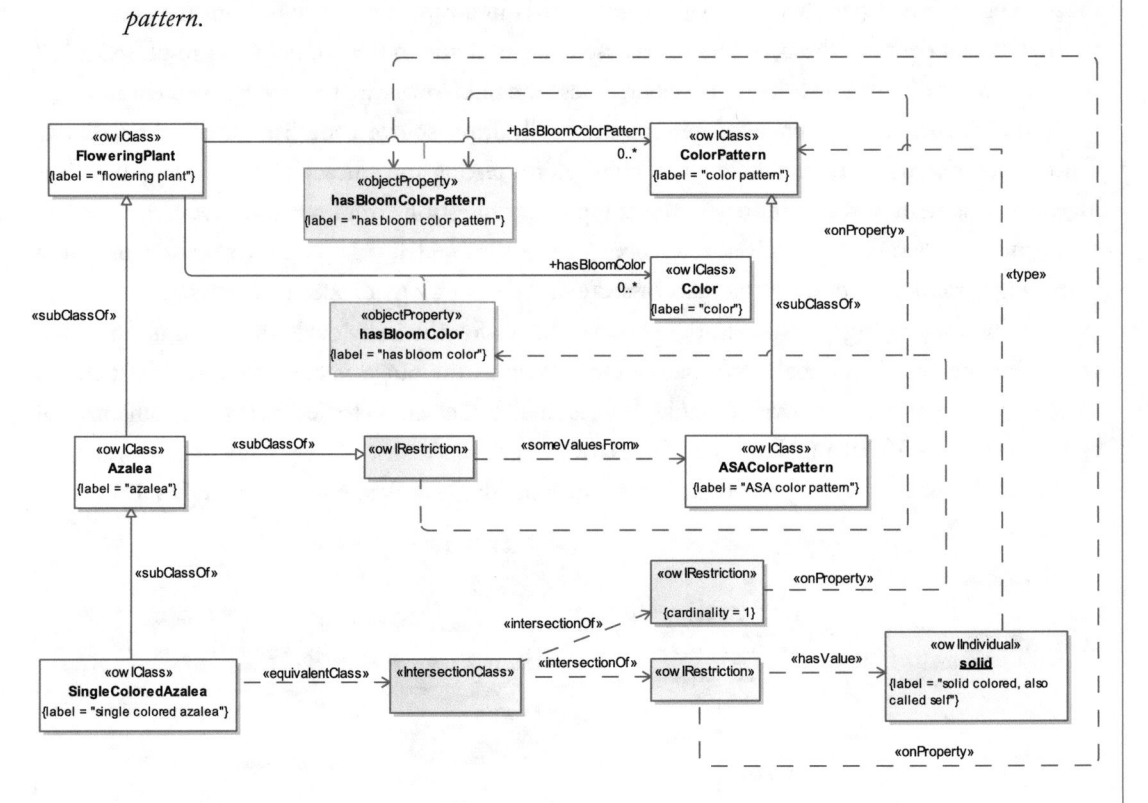

Figure 5.9: Example restricting some and exact values.

5.9 INDIVIDUALS AND DATA RANGES

An individual, also called an instance or object depending on the modeling paradigm, represents an object in the ontology. Individuals can be named or unnamed, but typically we give individuals explicit names so that they can be referenced by other ontologies.

- Any class that an individual is a member of, or is an individual of, is a type of the individual.

- Any superclass of a class that the individual is a member of is an ancestor of (or type of) the individual.

When we define individuals in an ontology, we typically restrict them to "*rarified*" or reference individuals, that are essential to describing the domain. That may include individuals representing the set of *ASAColorPatterns*, for example, that are essential to describing any azalea, or certain codes, such as the ISO 3166 country codes, used in many applications including but not limited to most websites. We recommend managing individuals in a separate ontology from the classes and properties that describe them, however, because changes made to the ontology can result in logical inconsistencies that can be difficult to debug unless the individuals are managed separately. In many cases, individuals can be generated programmatically from source code lists or other references. Only a representative set of individuals is required for testing the efficacy of an ontology (i.e., is it logically consistent and can it answer the competency questions from our use cases). A complete set of individuals reflecting a given reference, such as the NALT,[110] is typically managed only in a knowledge graph or data store that is accessed using the ontology to specify queries.

When specifying individuals, the property values for the individuals should conform to the constraints on the class (or classes) that define it, such as the range, value type, cardinality, etc. In cases where the number of "rarified" individuals is small but essential to describing the domain, such as the set of USDA Plant Hardiness Zones, or American Azalea Society color patterns, we use enumerated classes to specify the complete set of individuals or data ranges to specify valid values.

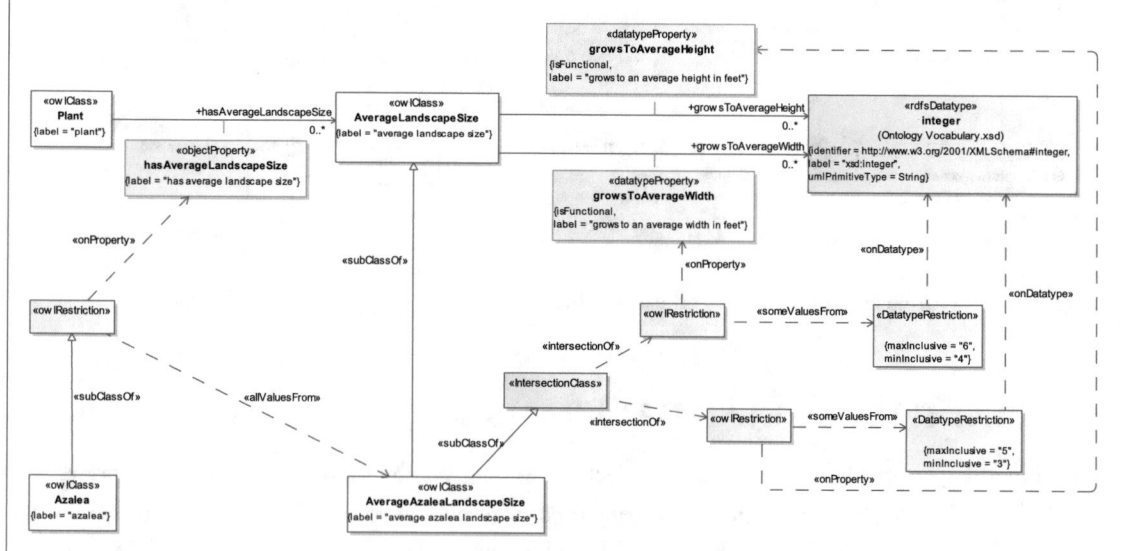

Figure 5.10: Example using complex data ranges and restrictions.

110 https://agclass.nal.usda.gov/.

Figure 5.10 provides a more complex example using data ranges and restrictions to answer the question: "*How large, on average, will an azalea grow to be at maturity?*"

5.10   OTHER COMMON CONSTRUCTS

Finally, there are a few additional constructs that we commonly use to further refine the semantics of a given concept. Two related property axioms that we use relatively frequently, especially for modeling and mapping specific data resources are the following.

- **Functional properties:** saying that a property is functional means that for every value in the domain there is exactly one value in the range.

- **Inverse functional properties:** modeling a property as inverse functional means that for every value in the range there is exactly one value in the domain, which is synonymous with the concept of a key in database terminology.

We use property chains to link properties together, often to simplify or improve search operations. We use disjointness and equivalences to enforce data governance rules, for data quality checking, for ontology consistency checking, and to uncover logic errors. Two classes are disjoint if they cannot have common instances. Disjoint classes cannot have any common subclasses, either. For example, if *winery* and *wine* are disjoint classes, then there is no instance that is both a winery and a wine; there is no class that is both a subclass of winery and a subclass of wine. In analyzing large ontologies, we may insert a disjointness relationship between two classes or properties where we think it makes sense to confirm that we really understand the ontology or to look for bugs. This strategy is especially useful when we get unexpected reasoning results and are attempting to understand why. Examples of unexpected results include (1) cases in which a class is inferred to be a subclass of something that it should not be and (2) cases in which an individual is inferred to be a member of a class that does not make sense. If we make classes explicitly disjoint, and an inference engine determines that the resulting ontology is logically inconsistent, then that tells us that either there is a logic error or something subtle is going on and perhaps additional clarification is required. If the resulting ontology is logically consistent but there are additional cases of unexpected classification, then again that indicates that the model likely requires clarification or revision in some way. Disjointness is also helpful in teasing out subtle distinctions among classes across multiple ontologies.

Similarly, equivalence expressions are also used frequently identify the same concepts across ontologies that may be named differently, or to name classes defined through class axioms, such as restrictions, if we want to reuse them.

CHAPTER 6

Conclusion

As we said at the start of Chapter 1, ontologies have become increasingly important as the use of knowledge graphs, machine learning, natural language processing (NLP), and the amount of data generated on a daily basis has exploded. As of 2014, 90% of the data in the digital universe was generated in the two years prior, and the volume of data was projected to grow from 3.2 zettabytes to 40 zettabytes in the next six years.[111] The very real issues that government, research, and commercial organizations are facing in order to sift through this amount of information to support decision-making alone mandate increasing automation. Yet, the data profiling, NLP, and learning algorithms that are ground-zero for data integration, manipulation, and search provide less than satisfactory results unless they utilize terms with unambiguous semantics, such as those found in ontologies and well-formed rule sets. There are a number of books available on the market today, many that are part of this series, on the nuts and bolts of the Resource Description Framework (RDF) and the Web Ontology Language (OWL). The gap that we observed when we started working on this book with respect to methodology and best practices for ontology engineering had not been addressed until now, however. This gap is really our motivation for writing this book. It is a work in progress, representing the material we give in the first 5–6 weeks of a 13-week course, but we intend to continue to update and add content, and keep it current as the field evolves.

The objectives of the course include:

- learning what ontologies are, how to build them, and how to use them in knowledge graphs and other settings;

- learning what use cases are, how to construct them, and how to use them to capture requirements for ontology and applications development;

- learning about terminology work, including definition development, and how to use terminologies as the starting point for ontology development;

- learning about ontology languages and some existing ontology resources; and

- learn how to design, implement, and evaluate an ontology-enabled project.

The content we've shared in this initial edition of the book covers the most fundamental building blocks necessary for any successful ontology project. In the first two chapters we provide background knowledge and essential reference material that reflect cross-disciplinary domains

---

[111] http://bigdata-madesimple.com/exciting-facts-and-findings-about-big-data/.

that overlap or converge in knowledge representation. Then in Chapter 3 we present an approach to requirements development and use case analysis that is critical for designing and scoping any ontology project. Most projects we've observed fail without a decent set of scoping requirements, and competency questions in particular. In Chapter 4 we provide an introduction to terminology work, which is a discipline in its own right, and often viewed as one of the most difficult tasks for many ontology projects. Developing good definitions, not even great definitions, can take a tremendous amount of work and is so important for many ontology projects, especially those that are part of an enterprise glossary or data governance initiative. Ontologies are often designed to enable machine *and human* communications and understanding, and it is the human side of the equation that many technologists struggle to address. Gaining consensus on terminology and definitions in a large organization can lead to many heated discussions, we know from personal experience, and yet is always enlightening for those participating in the work. We have seen countless "lightbulbs go on" for folks who thought they understood one another for years, and suddenly realized that they did not. In Chapter 5 we cover the starting points for building an ontology based on the use cases and competency questions, and using the terminology as a basis for that part of the task. This section is probably the one that is most often the starting point for many knowledge engineering tasks, that many papers and other books typically present (Noy and McGuinness, 2001; Borgida and Brachman, 2003).

But there is much more to talk about, including:

- best practices, validation and testing;

- information extraction and knowledge graph construction, including use of natural language processing;

- question answering;

- integrating provenance, references, and evidence citation for data management, governance, and lineage applications, among others, as well as to support explanation work;

- mapping strategies for integrating disparate data resources;

- using ontologies with machine learning, knowledge graph-based and rule-based applications; and

- ontology evaluation, evolution, and maintenance.

We cover these topics in our course, and our intent is to incorporate as much of this material as possible. We do not plan to cover all of the syntactic nuances or possible patterns in the use of the Resource Description Language (RDF), Web Ontology Language (OWL), or the SPARQL query language, all of which we use, as there are other references that do so (Allemang and Hendler, 2011;

Uschold M. , 2018). We will also limit coverage of description logic more generally, although there are best practices, validation methods, and other cases where we will touch on it. The best reference on description logic is *The Description Logic Handbook* (Baader et al., 2003), which we use in our course. We anticipate adding exercises as well as more worked examples over time.

We are always looking for feedback, and hope that our readers will provide that as well as requests for content that we might add in the future.

Bibliography

Allemang, D. and Hendler, J. (2011). *Semantic Web for the Working Ontologist: Effective Modeling in RDF and OWL* (Second Ed.). Waltham, MA: Morgan Kaufman Publishers. xvi, 94

Baader, F., Calvanese, D., McGuinness, D., Nardi, D., and Patel-Schneider, P. (2003). *The Description Logic Handbook: Theory, Implementation and Applications*. Cambridge, UK: Cambridge University Press. xv, 7, 65, 66, 84, 95

Bao, J., Kendall, E. F., McGuinness, D. L., and Patel-Schneider, P. F. (2012). *OWL 2. OWL 2 Web Ontology Language Quick Reference Guide* (Second Ed.). Cambridge, MA: World Wide Web Consortium (W3C). 7, 66

Beck, K. e. (2001). *Manifesto for Agile Software Development*. Agile Alliance. 35

Borgida, A. and Brachman, R. (2003). *Conceptual Modeling with Description Logics. The Description Logic Handbook: Theory, implementation, and applications*. Cambridge, UK: Cambridge University Press. 13, 94

Borgida, A., Brachman, R. J., McGuinness, D. L., and Resnick, L. A. (1989). CLASSIC: A structural data model for objects. *ACM SIGMOD International Conference on Management of Data* (pp. 59-67). Portland, OR: Association for Computing Machinery. DOI: 10.1145/67544.66932. 8

Brachman, R. J. and Levesque, H. J. (2004). *Knowledge Representation and Reasoning* (The Morgan Kaufmann Series in Artificial Intelligence). San Francisco, CA: Morgan Kaufmann Publishers, an Imprint of Elsevier, Inc. 2, 7

Brachman, R. J., Borgida, A., McGuinness, D. L., Patel-Schneider, P. F., and Resnick, L. A. (1991a). Living with CLASSIC: When and how to use a KL-ONE-Like language. In J. F. Sowa (Ed.), *Principles of Semantic Networks: Explorations in the Representation of Knowledge* (pp. 401–456). San Mateo, CA: Morgan Kaufman Publishers. DOI: 10.1016/B978-1-4832-0771-1.50022-9. xv

Brachman, R. J., McGuinness, D. L., Patel-Schneider, P. F., Resnick, L. A., and Borgida, A. (1991b). *Living with Classic. Principles of Semantic Networks*. (J. F. Sowa, Ed.) San Mateo, CA: Morgan Kaufman Publishers Inc. 401–456. DOI: 10.1016/B978-1-4832-0771-1.50022-9. 13, 66

Business Architecture Guild (2014). A Guide to the Business Architecture Body of Knowledge™, v. 4.1 (BIZBOK® GUIDE). 19

Chen, P. P.-S. (1976). The entity-relationship model: Toward a unified view of data. *ACM Transactions on Database Systems*, 1, 9–36. DOI: 10.1145/320434.320440. 14

Das, A., Wu, W., and McGuinness, D. L. (2001). Industrial strength ontology management. *Proceedings of the International Semantic Web Working Symposium*. Stanford, CA. 21

de Champeaux, D., Lea, D., and Faure, P. (1993). *Object-Oriented System Development*. Rahul, WI: Addison-Wesley. 13

Duque-Ramos, A., Fernández-Breis, J. T., Stevens, R., and Aussenac-Gilles, N. (2011). OQuaRE: A SQuaRE-based approach for evaluating the quality of ontologies. (R. V. García, Ed.) *Journal of Research and Practice in Information Technology*, 43(2), 41–58. 20

Green, R., Bean, C. A., and Myaeng, S. H. (Eds.). (2002). *The Semantics of Relationships: An Interdisciplinary Perspective*. Dordrecht, Netherlands: Springer Science+Business Media B.V. DOI: 10.1007/978-94-017-0073-3. 83

Grüninger, M. and Fox, M. S. (1995). Methodology for the design and evaluation of ontologies. *Proceedings, Workshop on Basic Ontological Issues in Knowledge Sharing*. Montreal: IJCAI-95. 38

Gruber, T. R. (1993). A translation approach to portable ontologies. *Knowledge Acquisition*, 5(2), 199–220. DOI: 10.1006/knac.1993.1008. 1

Guarino, N. and Musen, M. A. (2005). Applied ontology: Focusing on content. *Applied Ontology*, 1(1), 1–5. DOI: 10.3233/AO-150143.

Hart, L., Emery, P., Colomb, R., Chang, D., Raymond, K., Ye, Y., Kendall, E., and Dutra, M. (2004). Usage scenarios and goals for the ontology definition metamodel. In X. Xhou, S. Su, and M. Papazoglou (Eds.), *Web Information Systems Engineering Conference* (*WISE'04*). Brisbane, Australia: Springer LNCS. DOI: 10.1007/978-3-540-30480-7_62.

Hjørland, B. and Albrechtsen, H. (1995). Toward a new horizon in information science: Domain-analysis. *Journal of the American Society for Information Science*, 46(6), 400–425. DOI: 10.1002/(SICI)1097-4571(199507)46:6<400::AID-ASI2>3.0.CO;2-Y. 14

IDEF5 Ontology Description Capture Method. (1994). *Information Integration for Concurrent Engineering (IICE) IDEF5 Method Report*, 187. College Station, TX: Knowledge Based Systems, Inc. 54

ISO 1087-1: *Terminology Work – Vocabulary – Part 1: Theory and Application*. (2000). International Organization for Standardization (ISO). 45, 56, 62, 64

ISO 704:2000(E). (2000). Terminology Work – Principles and Methods. Geneva, Switzerland: ISO.

ISO 704:2009. (2009). *Terminology Work – Principles and Methods*. Geneva, Switzerland: ISO. 14, 54, 56

ISO/IEC 24707:2018 (2018). *Information Technology – Common Logic (CL)*. Geneva, Switzerland: International Organization for Standardization (ISO). 4, 7, 67, 84

ISO/IEC 24707:2007 (2007). *Information Technology – Common Logic (CL)*. Geneva, Switzerland: International Organization for Standardization (ISO).

Jacobson, I., Christerson, M., Jonsson, P., and Overgaard, G. (1992). *Object Oriented Software Engineering: A Use Case Driven Approach*. Boston, MA: Addison-Wesley Professional. 25, 35

Jacobson, I., Spence, I., and Bittner, K. (2011). *Use-Case 2.0: The Guide to Succeeding with Use Cases*. Sollentuna, Sweden: Ivar Jacobson International. 25, 26, 35

Ji, K., Wang, S., and Carlson, L. (2014). Matching and merging anonymous terms from web sources. *International Journal of Web and Semantic Technology* (*IJWesT*), 5(4). DOI: 10.5121/ijwest.2014.5404. 45

Kang, K. C., Cohen, S., Hess, J. A., Novak, W. E., and Peterson, A. S. (1990). *Feature-Oriented Domain Analysis (FODA) Feasibility Study*. Carnegie Mellon University (CMU), Software Engineering Institute (SEI). Pittsburgh, PA: CMU/SEI. DOI: 10.21236/ADA235785. 13

Klein, M. R. and Methlie, L. B. (1995). *Knowledge-Based Decision Support Systems with Applications in Business*. New York: John Wiley and Sons. 2

McDermott, D. and Doyle, J. (1980). Non-monotonic logic I. *Artificial Intelligence*, 13(1–2), 41-72. DOI: 10.1016/0004-3702(80)90012-0. 10

McGuinness, D. L. (2003). Ontologies come of age. In D. Fensel, J. Hendler, H. Lieberman, and W. Wahlster, (Eds.), *Bringing the World Wide Web to Its Full Potential*. Cambridge, MA: MIT Press. 3

McGuinness, D. L. and Wright, J. (1998). An industrial strength description logic-based configurator platform. *IEEE Intelligent Systems*, 13(4), 69–77. DOI: 10.1109/5254.708435. 8

McGuinness, D. L., Fikes, R., Rice, J., and Wilder, S. (2000). An environment for merging and testing large ontologies. *Proceedings of the Seventh International Conference on Principles of Knowledge Representation and Reasoning* (*KR2000*). Breckenridge, CO.

Nitsche, T., Mukerji, J., Reynolds, D., and Kendall, E. (2014). Using semantic web technologies for management application integration. *Semantic Web Enabled Software Engineering*, 17, 93–107. (J. Z. Pan and Y. Zhao, Eds.) Amsterdam, The Netherlands: IOS Press. 16, 65

Noy, N. F. and McGuinness, D. L. (2001). *Ontology Development 101: A Guide to Creating Your First Ontology*. Stanford University, Knowledge Systems Laboratory (KSL) and the Center for Biomedical Informatics Research. Stanford, CA: Stanford University. xv, 2, 66, 94

Ontology Definition Metamodel (ODM). *Version 1.1. (2014)*. Needham, MA: Object Management Group (OMG). 66

Patton, J. and Economy, P. (2014). *User Story Mapping: Discover the Whole Story, Build the Right Product*. Sebastopol, CA: O'Reilly Media Inc. 28

Sowa, J. F. (1999). *Knowledge Representation: Logical, Philosophical, and Computational Foundations*. Brooks Cole. 2, 7

Tartir, S., Arpinar, I. B., Moore, M., Sheth, A. P., and Aleman-Meza, B. (2005). *OntoQA: Metric-Based Ontology Quality Analysis*. Dayton, OH: The Ohio Center of Excellence in Knowledge-Enabled Computing (Kno.e.sis) Publications. 20. 21

Ulrich, W. and McWhorter, N. (2011). *Business Architecture: The Art and Practice of Business Transformation*. Tampa, FL: Meghan- Kiffer Press. 13

Unified Modeling Language (UML. *Version 2.5.1. (2017)*. Needham, MA: Object Management Group (OMG). 14, 66

Unified Modeling Language (UML). (2015). *Unified Modeling Language® (UML®)* 2.5. Needham, MA: Object Management Group (OMG).

Unified Modeling Language (UML). (2012). Needham, MA: Object Management Group (OMG). 35

Uschold, M. and Grüninger, M. (1996). Ontologies: Principles, methods, and applications. *Knowledge Engineering Review* (1), 96–137. DOI: 10.1017/S0269888900007797. 38

Uschold, M. (2018). Demystifying Owl for the enterprise. In Uschold, M. ,Ding, Y., and Groth, P. (Eds.), *Demystifying Owl for the Enterprise* (Synthesis Lectures on Semantic Web: Theory and Technology) (p. 237). Williston, VT: Morgan & Claypool. DOI: 10.2200/S00824ED1V01Y201801WBE017. 95

Wright, S. E. and Budin, G. (Eds.). (2001). *Handbook of Terminology Management: Volume 1: Basic Aspects of Terminology Management* (Vol. 1). Amsterdam, The Netherlands: John Benjamins Publishing Company. DOI: 10.1075/z.htm2. 45

Yao, H., Orme, A. M., and Etzkorn, L. (2005). Cohesion metrics for ontology design and application. *Journal of Computer Science*, 1(1) , 107–113. DOI: 10.3844/jcssp.2005.107.113. 21

Author's Biographies

Elisa F. Kendall is a Partner in Thematix Partners LLC and graduate-level lecturer in computer science, focused on data management, data governance, knowledge representation, and decisioning systems. Her consulting practice includes business and information architecture, knowledge representation strategies, and ontology design, development, and training for clients in financial services, government, manufacturing, media, pharmaceutical, and retail domains. Recent projects have focused on use of ontologies to drive natural language processing, machine learning, interoperability, and other knowledge graph-based applications. Elisa represents knowledge representation, ontology, information architecture, and data management concerns on the Object Management Group (OMG)'s Architecture Board, is co-editor of the Ontology Definition Metamodel (ODM), and a contributor to a number of other ISO, W3C, and OMG standards, including the Financial Industry Business Ontology (FIBO) effort. Prior to joining Thematix, she was the founder and CEO of Sandpiper Software, an early entrant in the Semantic Web domain. Earlier in her career, she was software development manager for Aspect Development, and before that a ground systems data processing engineer for Lockheed Martin. She holds a B.S. in Mathematics and Computer Science from UCLA, and an A.M in Linguistics from Stanford University.

Deborah L. McGuinness is the Tetherless World Senior Constellation Chair and Professor of Computer, Cognitive, and Web Sciences at RPI. She is also the founding director of the Web Science Research Center and the CEO of McGuinness Associates Consulting. Deborah has been recognized with awards as a fellow of the American Association for the Advancement of Science (AAAS) for contributions to the Semantic Web, knowledge representation, and reasoning environments and as the recipient of the Robert Engelmore award from the Association for the Advancement of Artificial Intelligence (AAAI) for leadership in Semantic Web research and in bridging Artificial Intelligence (AI) and eScience, significant contributions to deployed AI applications, and extensive service to the AI community. Deborah leads a number of large diverse data intensive resource efforts and her team is creating next-generation ontology-enabled research infrastructure for work in large interdisciplinary settings. Prior to joining RPI, Deborah was the acting director of the Knowledge Systems, Artificial Intelligence Laboratory, and Senior Research Scientist in the Computer Science Department of Stanford University, and previous to that she was at AT&T Bell Laboratories. Deborah consults with numerous large corporations as well as emerging startup companies wishing to plan, develop, deploy, and maintain Semantic Web and/or AI applications. Some areas of recent work include: data science, next-generation health advisors, ontology design and evolution environments, semantically enabled virtual observatories, semantic integration of scientific data, context-aware mobile applications, search, eCommerce, configuration, and supply chain management. Deborah holds a Bachelor of Math and Computer Science from Duke University, a Master of Computer Science from University of California at Berkeley, and a Ph.D. in Computer Science from Rutgers University.

Printed in the United States
by Baker & Taylor Publisher Services